β Beta

Focus: Multiple-Digit
Addition and Subtraction

Instruction Manual

By Steven P. Demme

1-888-854-MATH (6284)
www.MathUSee.com

Math·U·See

1-888-854-MATH (6284)
www.MathUSee.com

Math·U·See

SCOPE & SEQUENCE

Math-U-See is a complete and comprehensive K-12 math curriculum. While each book focuses on a specific theme, Math-U-See continuously reviews and integrates topics and concepts presented in previous levels.

Primer

α Alpha | Focus: Single-Digit Addition and Subtraction

β Beta | Focus: Multiple-Digit Addition and Subtraction

γ Gamma | Focus: Multiplication

δ Delta | Focus: Division

ε Epsilon | Focus: Fractions

ζ Zeta | Focus: Decimals and Percents

Pre-Algebra

Algebra 1

Stewardship*

Geometry

Algebra 2

Pre Calculus with Trigonometry

Stewardship is a biblical approach to personal finance. The requisite knowledge for this curriculum is a mastery of the four basic operations, as well as fractions, decimals, and percents. In the Math-U-See sequence these topics are thoroughly covered in Alpha through Zeta. We also recommend Pre-Algebra and Algebra 1 since over half of the lessons require some knowledge of algebra. Stewardship may be studied as a one-year math course or in conjunction with any of the secondary math levels.

HOW TO USE

Five Minutes for Success

Welcome to Beta. I believe you will have a positive experience with the unique Math-U-See approach to teaching math. These first few pages explain the essence of this methodology which has worked for thousands of students and teachers. I hope you will take five minutes and read through these steps carefully.

I am assuming your student has mastered the addition and subtraction facts.

If you are using the program properly and still need additional help, you may contact your authorized representative, or visit Math-U-See online at http://www.mathusee.com/support.html

— S. Demme

The Goal of Math-U-See

The underlying assumption or premise of Math-U-See is that the reason we study math is to apply math in everyday situations. Our goal is to help produce confident problem solvers who enjoy the study of math. These are students who learn their math facts, rules, and formulas and are able to use this knowledge to solve word problems and real life applications. Therefore, the study of math is much more than simply committing to memory a list of facts. It includes memorization, but it also encompasses learning the underlying concepts of math that are critical to successful problem solving.

More than Memorization

Many people confuse memorization with understanding. Once while I was teaching seven junior high students, I asked how many pieces they would each receive if there were fourteen pieces. The students' response was, "What do we do: add, subtract, multiply, or divide?" Knowing **how** to divide is important, understanding **when** to divide is equally important.

THE SUGGESTED 4-STEP MATH-U-SEE APPROACH

In order to train students to be confident problem solvers, here are the four steps that I suggest you use to get the most from the Math-U-See curriculum.

Step 1. Prepare for the Lesson
Step 2. Present the New Topic
Step 3. Practice for Mastery
Step 4. Progression after Mastery

Step 1. Prepare for the Lesson.

Watch the DVD to learn the new concept and see how to demonstrate this concept with the manipulatives when applicable. Study the written explanations and examples in the instruction manual. Many students watch the DVD along with their instructor.

Step 2. Present the New Topic

Present the new concept to your student. Have the student watch the DVD with you, if you think it would be helpful. Older students may watch the DVD on their own.

a. **Build:** Use the manipulatives to demonstrate the problems from the worksheet.

b. **Write:** Record the step-by-step solutions on paper as you work them through with manipulatives.

c. **Say:** Explain the *why* and *what* of math as you build and write.

Do as many problems as you feel are necessary until the student is comfortable with the new material. One of the joys of teaching is hearing a student say *"Now I get it!"* or *"Now I see it!"*

Step 3. Practice for Mastery.

Using the examples and the lesson practice problems from the student text, have the students practice the new concept until they understand it. It is one thing for students to watch someone else do a problem, it is quite another to do the same problem themselves. Do enough examples together until they can do them without assistance.

Do as many of the lesson practice pages as necessary (not all pages may be needed) until the students remember the new material and gain understanding. Give special attention to the word problems, which are designed to apply the concept being taught in the lesson.

Another resource is the Math-U-See web site which has online drill and downloadable worksheets for more practice. Go to www.mathusee.com and select "Online Helps."

Step 4. Progression after Mastery.

Once mastery of the new concept is demonstrated, proceed to the systematic review pages for that lesson. Mastery can be demonstrated by having each student teach the new material back to you. The goal is not to fill in worksheets, but to be able to teach back what has been learned.

The systematic review worksheets review the new material as well as provide practice of the math concepts previously studied. Remediate missed problems as they arise to ensure continued mastery.

Proceed to the lesson tests. These were designed to be an assessment tool to help determine mastery, but they may also be used as extra worksheets. Your students will be ready for the next lesson only after demonstrating mastery of the new concept and continued mastery of concepts found in the systematic review worksheets.

Confucius was reputed to have said, "Tell me, I forget; Show me, I understand; Let me do it, I will remember." To which we add, **"Let me teach it and I will have achieved mastery!"**

Length of a Lesson

So how long should a lesson take? This will vary from student to student and from topic to topic. You may spend a day on a new topic, or you may spend several days. There are so many factors that influence this process that it is impossible to predict the length of time from one lesson to another. I have spent three days on a lesson and I have also invested three weeks in a lesson. This occurred in the same book with the same student. If you move from lesson to lesson too quickly without the student demonstrating mastery, he will become overwhelmed and discouraged as he is exposed to more new material without having learned the previous topics. But if you move too slowly, your student may become bored and lose interest in math. But I believe that as you regularly spend time working along with your student, you will sense when is the right time to take the lesson test and progress through the book.

By following the four steps outlined above, you will have a much greater opportunity to succeed. Math must be taught sequentially, as it builds line upon line and precept upon precept on previously learned material. I hope you will try this methodology and move at your student's pace. As you do, I think you will be helping to create a confident problem solver who enjoys the study of math.

ONGOING SUPPORT
AND ADDITIONAL RESOURCES

Welcome to the Math-U-See Family!

Now that you have invested in your children's education, I would like to tell you about the resources that are available to you. Allow me to introduce you to your regional representative, our ever improving website, the Math-U-See blog, our new free e-mail newsletter, the online Forum, and the Users Group.

Most of our regional **Representatives** have been with us for over 10 years. What makes them unique is their desire to serve and their expertise. They have all used Math-U-See and are able to answer most of your questions, place your student(s) in the appropriate level, and provide knowledgeable support throughout the school year. They are wonderful!

Come to your local curriculum fair where you can meet your rep face-to-face, see the latest products, attend a workshop, meet other MUS users at the booth, and be refreshed. We are at most curriculum fairs and events. To find the fair nearest you, click on "Events Calendar" under "News."

The **Website**, at www.mathusee.com, is continually being updated and improved. It has many excellent tools to enhance your teaching and provide more practice for your student(s).

ONLINE DRILL

Let your students review their math facts online. Just enter the facts you want to learn and start drilling. This is a great way to commit those facts to memory.

WORKSHEET GENERATOR

Create custom worksheets to print out and use with your students. It's easy to use and gives you the flexibility to focus on a specific lesson. Best of all — it's free!

Math-U-See Blog

Interesting insights and up-to-date information appear regularly on the Math-U-See Blog. The blog features updates, rep highlights, fun pictures, and stories from other users. Visit us and get the latest scoop on what is happening .

Email Newsletter

For the latest news and practical teaching tips, sign up online for the free Math-U-See e-mail newsletter. Each month you will receive an e-mail with a teaching tip from Steve as well as the latest news from the website. It's short, beneficial, and fun. Sign up today!

The Math-U-See Forum and the Users Group put the combined wisdom of several thousand of your peers with years of teaching experience at your disposal.

Online Forum

Have a question, a great idea, or just want to chitchat with other Math-U-See users? Go to the online forum. You can also use the forum to post a specific math question if you are having difficulty in a certain lesson. Head on over to the forum and join in the discussion.

Yahoo Users Group

The MUS-users group was started in 1998 for lovers and users of the Math-U-See program. It was founded by two home-educating mothers and users of Math-U-See. The backbone of information and support is provided by several thousand fellow MUS users.

For Specific Math Help

When you have watched the DVD instruction and read the instruction manual and still have a question, we are here to help. Call your local rep, click the support link and e-mail us here at the home office, or post your question on the forum. Our trained staff have used Math-U-See themselves and are available to answer a question or walk you through a specific lesson.

Feedback

Send us an e-mail by clicking the feedback link. We are here to serve you and help you teach math. Ask a question, leave a comment, or tell us how you and your student are doing with Math-U-See.

Our hope and prayer is that you and your students will be equipped to have a successful experience with math!

Blessings,

Steve Demme

LESSON 1

Place Value and the Manipulatives

Two skills are needed to function in the decimal system: the ability to count from zero to nine, and an understanding of place value. In the decimal system, where everything is based on ten (deci), you count to nine and then start over. To illustrate this, count the following numbers slowly: 800, 900, 1000. We read these as eight hundred, nine hundred, one thousand. Now read these: 80, 90, 100. Notice how you count from one to nine and then begin again. These are read as eighty, ninety, one hundred. Once you can count to nine, then begin work on place value. The two keys are learning the counting numbers zero through nine, which tell us how many; and understanding place value, which tells us what kind.

Counting

When counting, begin with zero, then proceed to nine. Traditionally, we've started with one and counted to ten. Look at the two charts that follow and see which is more logical.

1	2	3	4	5	6	7	8	9	10		0	1	2	3	4	5	6	7	8	9
11	12	13	14	15	16	17	18	19	20		10	11	12	13	14	15	16	17	18	19
21	22	23	24	25	26	27	28	29	30		20	21	22	23	24	25	26	27	28	29

The second chart has all single digits in the first line. Then in the second line, each number is preceded by a one in the tens place. In the next line, a two precedes each number instead of a one. The first chart, though looking more familiar, has the 10, the 20, and the 30 in the wrong lines. When counting, always begin with zero and count to nine, and then start over.

When explaining this important subject, I tell students, "Every value has its own place!" To an older child I would add, "Place determines value!" Both are true. There are ten symbols to tell you how many, and many values to represent what kind or what value. Zero through nine tell us how many; units, tens, and hundreds tell us what kind. For the sake of accuracy, *units* will be the word used to denote the first value, instead of ones. One is a counting number that tells us how many, and units is a place value that denotes what kind. This will save potential confusion when saying ten ones or one ten. Remember, one is a number and units is a place value. The numerals 0–9 tell us how many tens, how many hundreds, or how many units. We begin our study focusing on the units, tens, and hundreds, but there are other values such as thousands, millions, billions, and so on.

To illustrate this lesson, I like to use a street, since I'm talking about a place. I call the street Decimal Street. On this street I have the little green units house, the tall blue tens house next door, and the huge red hundreds castle next to the tens. We don't want to forget what we learned from counting—that we count only to nine and then start over. To make this more real, begin by asking, "What is the largest number of units that can live in this house?" You can get any response to this question, from zero to nine, and you might say "yes" to all of them, but remind the student that the largest number is nine! So we imagine how many little green beds, or green toothbrushes, or green chairs there would be in the house. Ask the student what else there might be nine of. Do the same with the tens and the hundreds. Remember that in these houses all the furniture will be blue (tens) and red (hundreds).

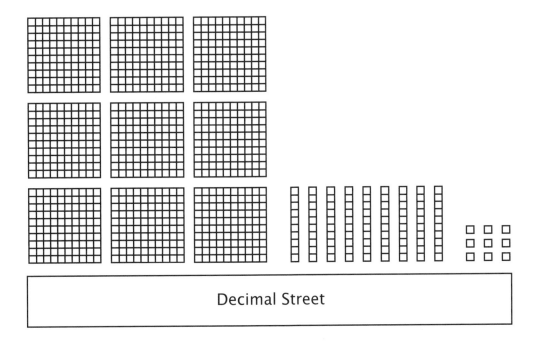

Decimal Street

Throughout the program, whenever we teach we will employ the following strategy: Build, Write, Verbalize. To teach place value, we will first build the number, then count how many are in each place, then write the number and read what we've written.

Let's build 142 (1 hundred, 4 tens, 2 units). Then count how many are "home" at each house. I like to imagine going up to the door of each home and knocking to see how many are home in each place. Then write the numerals 1 4 2 as you count (always beginning with the units) to show the value on paper. Then say, "One hundred four tens and two units, or one hundred forty-two." Build another number and have the student write how many are home. When they understand this, then you write how many on paper and have them build it! Try 217. After they build it, read what they have built. Keep practicing, going back and forth between the teacher building and the student writing, and vice versa.

Here is another exercise I do to reinforce the fact that every value has its own place. I like to have the student close his/her eyes as I move the pieces around, placing the red hundreds where the units should be, and vice versa. I then ask the student to make sure they are all in the right place. You might call this "scramble the values" or "walk the blocks home." As the student looks at the problem and begins to work on it, I ask, "Is every value in its own place?"

An important symbol that I haven't mentioned is Mr. Zero. He is a place holder. Let's say you were walking down Decimal Street and knocking on each door to see who was home. If you knock on the units door and three units answer, then you have three in the units place. Next door at the tens house, you knock and no one answers. Yet you know someone is there feeding the goldfish and taking care of the bird, as someone might do when a family goes on vacation. He won't answer the door because he's not a ten, but he's the one who holds their place until the tens come home. Upon knocking at the big red hundreds castle, you find that two hundreds answer the door. Your numeral is thus: 203.

At some point you will want to mention that even though we begin at the units end of the street and proceed right to left, from the units to the hundreds, when we read the number we do it left to right. We want to get into the habit of counting units first, so when we add, we will add units first, then tens, then hundreds. On the homework sheets have the student count the correct number at the top of the page, then have them build the number shown next.

Remember we teach with the blocks, and then we move to the worksheets once the student understands the new material.

You've probably noticed the important relationship between language and place value. Consider 142, read as "one hundred forty-two." We know that it is made up of one red hundred square (one hundred) and four blue ten bars (forty–*ty* for ten) and two units. The hundreds are very clear and self-explanatory, but the tens are where we need to focus our attention.

When pronouncing 90, 80, 70, 60, and 40, work on enunciating clearly so that 90 is ninety, not "ninedee." 80 is eighty, not "adee." When you pronounce the numbers accurately, not only will your spelling improve, but your understanding of place value as well. Seventy (70), is seven tens and sixty (60), is six tens. Forty (40) is pronounced correctly but spelled without the *u*. Carrying through on this logic, 50 should be pronounced "five-ty" instead of fifty. Thirty and twenty are similar to fifty, not completely consistent but close enough so we know what they mean. The teens are the real problem

Some researchers have concluded that one of the chief differences between Western (American and Canadian) and Eastern (Chinese and Japanese) students is their understanding of place value. The culprit, in the researcher's eyes, is the English language. In Eastern languages, when a child can count to nine, with a few minor variations they can count to one hundred. Not so in English, with such numbers as ten, eleven, twelve and the rest of the teens. Not only are these numbers difficult to teach, because there seems to be no rhyme nor reason as to their origin, but more importantly, they do not reflect and indicate place value. To remedy this serious deficiency, I'm suggesting a new way to read the numbers 10 through 19. You decide whether this method reinforces the place value concept and restores logic and order to the decimal system. Ten is "onety," 11 is "onety-one," 12 is "onety-two," 13 is "onety-three," and so on. Now, it is not that students can't say ten, eleven, twelve, but learning this method enhances their understanding, makes math logical again, and they think it is neat.

When presenting place value or any other topic in this curriculum, model how you think as you solve the problems. As you, the teacher, work through a problem with the manipulatives, do it verbally, so that as the student observes, he/she also hears your thinking process. Then record your answer.

Example 1

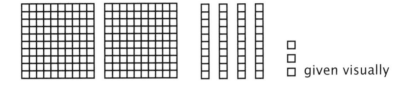

given visually

As you look at the picture, say it slowly, proceeding from left to right: "two hundred forty-three." Then count, beginning with the units, "1-2-3," and write a 3 in the units place. Then count the tens, "1-2-3-4," and write a 4 in the tens place. Finally, count the hundreds, "1-2," and write a 2 in the hundreds place. Do several of these, then give the student the opportunity to do some.

Example 2

274 (given as a written number)

Read the number, "two hundred seventy-four," then say "two hundreds" as you pick up two red hundred squares. Then say "seven-ty or seven tens" and pick up seven blue ten bars. Then say "four" and pick up four green unit pieces. Then place the two hundreds in the correct place as you say, "Every value has its own place." Do several of these, then give the student the opportunity to do some.

Example 3

"one hundred sixty-five" (given verbally)

Read the number slowly, then say "one hundred" as you pick up one red hundred square. Then say "six-ty or six tens" and pick up six blue ten bars. Then say "five" and pick up five green unit pieces. Then place them in the correct place as you say, "Every value has its own place." Then write the number 165. Do several of these, then give the student the opportunity to do some.

Game—Pick a Card

Make up a set of cards with 0 through 9 written in green, with one number on each card. Then make another stack of cards with the same numbers written in blue. Create one more stack of cards with the numbers 0 through 9 written in red. Shuffle the green cards, pick one, and display that number of green unit blocks. If a child picks a green 4, then count out four green unit blocks and show them.

When the child is proficient at this game, try it with the blue cards and do the same thing except choose the blue ten blocks instead of the green unit blocks. When they can do the tens well, combine the green cards and the blue cards. Have the child choose one card from the green pile and one card from the blue pile and then pick up the correct number of blue ten blocks and green unit blocks. When they are experts at this, add the red cards and proceed as before. Place them in three stacks, shuffle, and draw from each stack. Have the student show you with the blocks what number they have drawn.

Get a large piece of paper for your background, and cut out three houses. The units house should be green and 1/2" x 4 1/2" (as a rectangle) or 1 1/2" x 1 1/2" (as a square). The tens house should be blue and 4 1/2" x 5". The hundreds house should be red and 15" x 15". Each of these should be able to hold exactly nine of each piece. This drawing is not to scale.

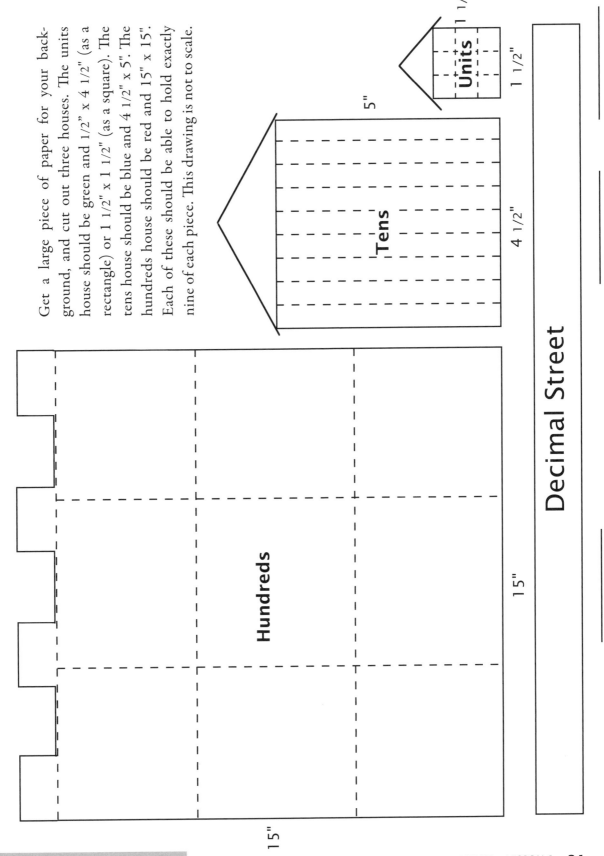

Units

1 1/2"

1 1/2"

Tens

5"

4 1/2"

Hundreds

15"

15"

Decimal Street

Sequencing; Word Problem Tips

Sequencing is putting numbers in order from the smallest to the largest or the largest to the smallest. Using the blocks to represent the numbers before putting them in order makes this exercise very clear. Remember to build, write, and say when doing these problems. The next lesson uses the words and symbols for greater than and less than, while this lesson helps get the student ready to compare numbers by asking the questions, "Which is smaller?" and "Which is larger?"

This topic also reinforces and reviews place value, which is the cornerstone of the decimal system. Comparing numbers in the same place value is not very difficult. But when comparing and sequencing numbers with different place values, it is imperative to use the blocks to show why the sequence behaves as it does. For example, when comparing numbers such as 95 and 123 it might be helpful to put a 0 in the hundreds place so you are comparing 095 and 123. When comparing numbers with different place values like this, start with the largest value, in this case the hundreds, and then move to the right. We know that the larger values are to the left, so start from the largest place value and move to the smallest, or from left to right. Emphasize to the student that sequencing is a two-step process. First you have to compare the numbers, then you have to put them in order. Study the following examples carefully.

When they get proficient at sequencing, we'll do some problems with filling in the missing numbers. (See examples 5 and 6.) The student may use the blocks for this as well. However, the examples just use blank spaces and numbers.

Example 1

Put the numbers in order from the smallest to the largest.

8, 3, 6

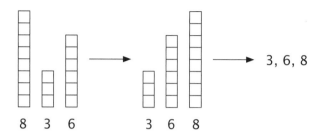

3, 6, 8

Example 2

Put the numbers in order from the smallest to the largest.

10, 50, 20

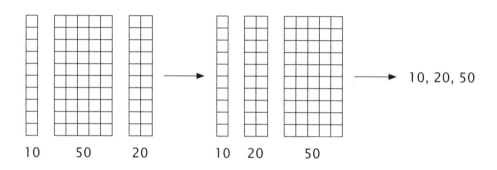

10, 20, 50

Example 3

Put the numbers in order from the largest to the smallest.

14, 23, 9

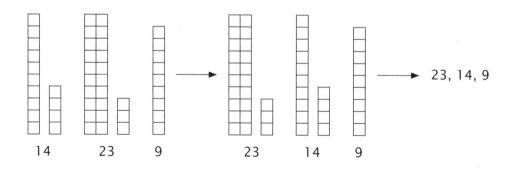

23, 14, 9

Example 4

Put the numbers in order from the largest to the smallest.

49, 103, 61

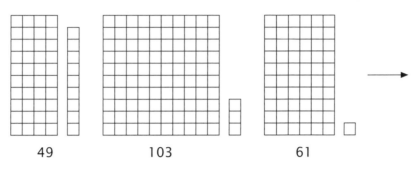

49   103   61

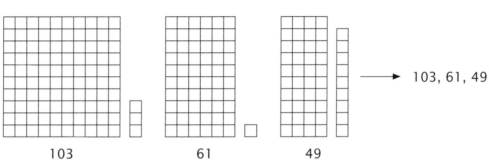

103   61   49   103, 61, 49

Example 5

Fill in the blank spaces with the correct numbers for the sequence.

___9___ , _____ , ___7___ , ___6___ , _____

Solution ___9___ , ___8___ , ___7___ , ___6___ , ___5___

Example 6

Fill in the blank spaces with the correct numbers for the sequence.

_____ , ___3___ , ___4___ , _____ , _____

Solution ___2___ , ___3___ , ___4___ , ___5___ , ___6___

Word Problem Tips

Parents often find it challenging to teach children how to solve word problems. Here are some suggestions for helping your student learn this important skill.

The first step is to realize that word problems require both reading and math comprehension. Don't expect a child to be able to solve a word problem if he does not thoroughly understand the math concepts involved. On the other hand, a student may have a math skill level that is stronger than his or her reading comprehension skills. Below are a number of strategies to improve comprehension skills in the context of story problems. You may decide which ones work best for you and your child.

Strategies for word problems

1. Ignore numbers at first and read the story. It may help some students to read the question aloud. Every word problem tells a story. Before deciding what math operation is required, let the student retell the story in his own words. Who is involved? Are they receiving gifts, losing something, or dividing a treat?

2. Relate the story to real life, perhaps by using names of family members. For some students, this makes the problem more interesting and relevant.

3. Build, draw, or act out the story. Use the blocks or actual objects when practical. Especially in the lower levels, you may require the student to use the blocks for word problems even when the facts have been learned. Don't be afraid to use a little drama as well. The purpose is to make it as real and meaningful as possible.

4. Look for the common language used in a particular kind of problem. Pay close attention to the word problems on the lesson practice pages, as they model different kinds of language that may be used for the new concept just studied. For example, "altogether" indicates addition. These "key words" can be useful clues but should not be a substitute for understanding.

5. Look for practical applications that use the concept and ask questions in that context.

6. Have the student invent word problems to illustrate their number problems from the lesson.

Cautions

1. Unneeded information may be included in the problem. For example, we may be told that Suzie is eight years old, but the eight is irrelevant when adding up the number of gifts she received.

2. Some problems may require more than one step to solve. Model these questions carefully.

3. There may be more than one way to solve some problems. Experience will help the student choose the easier or preferred method.

4. Estimation is a valuable tool for checking an answer. If an answer is unreasonable, it is possible that the wrong method was used to solve the problem.

Inequalities

Read 2 < 3 as "two is less than three." Read 4 > 1 as "four is greater than one." The small end of the symbol always points to the smaller number. The open or larger end points to the larger value. Some say the symbol represents an alligator's mouth; the alligator always wants to eat the larger number. Invent your own saying! Put the <, >, or = sign in the oval shape to make the expression true.

Example 1

12 ◯ 8

12 is greater than 8, so the symbol is >.   12 ⟨>⟩ 8

Example 2

7 ◯ 9

7 is less than 9, so the symbol is <.   7 ⟨<⟩ 9

Example 3

4 ◯ 4

4 is equal to 4, so the symbol is =.   4 ⟨=⟩ 4

Example 4

5 + 7 ◯ 9 + 4

12 ◯ 13   Add both sides.

12 is less than 13, so the symbol is <.   12 ⟨<⟩ 13

Example 5

10 - 3 ◯ 9 - 4

7 ◯ 5   Subtract both sides.

7 is greater than 5, so the symbol is >.   7 ⟩ 5

Game—Fill in the Box

Get a large piece of white paper and draw a vertical line down the middle. Then add a box in the center of the dividing line. Make cards with >, <, and =. Put different numbers of blocks on each side of the line, and then choose the appropriate card to place in the box.

Facts Review

We will be reviewing addition and subtraction facts throughout the student textbook. If you find that you need more review of these facts, consult the Math-U-See Web site, which provides online drill and downloadable worksheets. Go to *www.mathusee.com* and click on Online Helps.

Rounding to 10 and Estimation

Rounding to 10 is used in estimating before we add. When you *round* a number to the nearest multiple of 10, there will be a number in the tens place, but only a zero in the units place. I tell the students this is why we call it rounding, because the units are going to be a "round" zero.

Let's round 38 as an example. The first skill is to find the two multiples of 10 that are nearest to 38. The lower one is 30 and the higher one is 40. Thirty-eight is between 30 and 40. If the student has trouble finding these numbers, begin by placing your finger over the 8 in the units place, and all you have is a 3 in the tens place, which is 30. Then add one more to the tens to find the 40. I often write the numbers 30 and 40 above the number 38 on both sides, as in figure 1.

Figure 1   30 ₃₈ 40

The next skill is to find out if 38 is closer to 30 or 40. Let's go through all the numbers as given in figure 2. It is obvious that 31, 32, 33, and 34 are closer to 30, and 36, 37, 38, and 39, are closer to 40. But 35 is right in the middle. Somebody decided that it goes to 40, so that is the reason for our rule. When rounding to tens, look at the units place. If the units are 0, 1, 2, 3, or 4, the digit in the tens place remains unchanged. If the units are 5, 6, 7, 8, or 9, the digit in the tens place increases by one. See figure 2.

Figure 2

30   31   32   33   34   35   36   37   38   39   40

Another strategy I use is to put 0, 1, 2, 3, and 4 inside a circle to represent zero, because if these numbers are in the units place, they add nothing to the tens place. This means they are rounded to the lower number (30 in the example). Then I put 5, 6, 7, 8, and 9 inside a thin rectangle to represent one, because if these numbers are in the units place, they do add one to the tens place. This means they are rounded to the higher number (40 in the example).

Figure 3

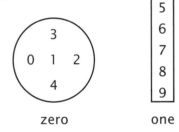

zero   one

Example 1

Round 13 to the nearest tens place.

10   20   1. Find the multiples of 10 nearest to 13.
   13

10   20   2. We know that 3 goes to the lower number, 10.
   13

(10) 20   3. Or, recall that 3 is in the circle, or zero, so nothing
   13   is added to the smaller number, 10.

Example 2

Round 75 to the nearest tens place.

70   80   1. Find the multiples of 10 nearest to 75.
   75

70   80   2. We know that 5 goes to the higher number, 80.
   75

70   (80)   3. Recall that 5 is in the rectangle, or one, so 1 is
   75   added to the 7, making the answer 80.

Estimation

Now that we have learned to round numbers, we can apply this knowledge to helping us *estimate*, or find the approximate answer. In the next lesson, we will compare the estimate to the real answer, and if the numbers are close, we are probably correct in our computation.

Example 3

$$
\begin{array}{r}
13 \quad (10) \\
+75 \quad (80) \\
\hline
88 \quad (90)
\end{array}
$$

Round the numbers and put them in parentheses.
Then add the estimates to get 90.

After working the problem exactly, compare your
answers to see if they are close.

Commutative Squares

A fun way to review addition facts is to use what some call magic squares. You add the numbers vertically to find two sums (#1), then add the numbers horizontally to find two more sums (#2). After doing this, you add the sums at the bottom horizontally to find the sum that goes in the bottom right square (#3). Finally, add the sums on the right vertically to find the answer for the bottom right-hand square (#4). The results of the last two problems should agree in the bottom right square. If not, some of your addition needs to be corrected. Have fun with the examples in the student text.

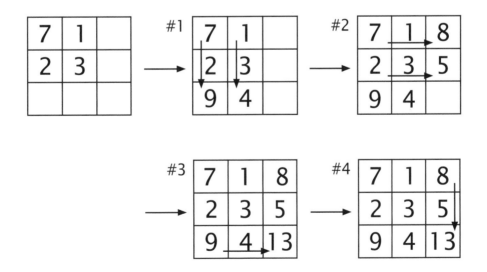

Multiple-Digit Addition
Place-Value Notation

Add the units first. Always add from right to left, from smaller to larger. If this is confusing, simply explain that we read the numbers from left to right, but we combine them from right to left. Even though it may not make a difference at this point, begin properly. Place the four bar and the five bar end to end and place a nine bar next to, or on top of, it to show that 4 + 5 = 9. Next, push all the tens together so you see that two tens plus three tens is the same as five tens. Then add the hundreds. Notice that you add units to units, tens to tens, and hundreds to hundreds. Whenever you add two numbers, you always add the same values. To combine, numbers must be the same kind. Here are three examples followed by the same problems worked out with place value notation.

Place-Value Notation

Place-value notation is simply writing out the numbers and separating the place values. It follows the format of the blocks. For example, 123 is written as 100 + 20 + 3. This notation reinforces place value.

Example 1
Write 971 with place-value notation.

900 + 70 + 1

Example 2

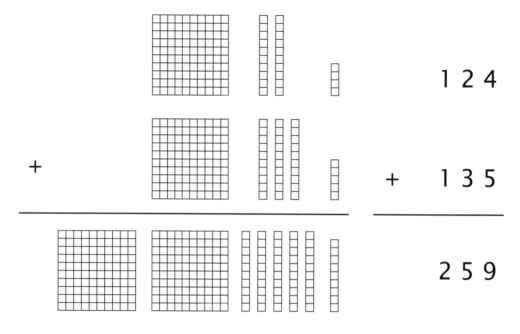

place-value notation

$$124 = 100 + 20 + 4$$
$$+135 = 100 + 30 + 5$$
$$259 = 200 + 50 + 9$$

The numbers being added are called the *addends*. The answer is the *sum*. In the example above, 124 and 135 are addends, and 259 is the sum.

Example 3

$$1 \; 0 \; 6$$

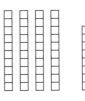

$$+ \quad \; 4 \; 2$$

$$1 \; 4 \; 8$$

place-value notation

```
106=100   +6
+ 42=     40+2
148=100+40+8
```

Example 4

$$2 \; 3 \; 0$$

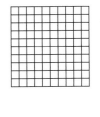

$$+1 \; 4 \; 8$$

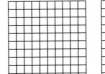

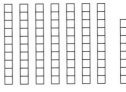

$$3 \; 7 \; 8$$

place-value notation

```
230=200+30
+148=100+40+8
378=300+70+8
```

Skip Count by 2

Skip counting is the ability to count groups of the same number quickly. For example, if you were to skip count by three you would skip the 1 and the 2 and say "3," then skip the 4 and the 5 and say "6," then "9 - 12 - 15 - 18," etc. Skip counting by seven is 7 - 14 - 21 - 28 - 35 - 42 - 49 - 56 - 63 - 70.

Here are five reasons for learning to skip count.

1. This skill lays a solid foundation for learning your multiplication facts. We can write 3 + 3 + 3 + 3 as 3 x 4. If a child can skip count, he could say, "3 - 6 - 9 - 12." Then he could read 3 x 4 as "3 counted 4 times is 12." As you learn your skip-counting facts, you are learning all of the products in the multiplication facts in order. Multiplication is fast adding of the same number, and skip counting illustrates this beautifully. You can think of multiplication as a shortcut to the skip-counting process. Consider 3 x 5. I could skip count 3 five times as 3 - 6 - 9 - 12 - 15 to come up with the solution. Or after I learn my facts I can say, "3 counted 5 times is 15." The latter is much faster.

2. Skip counting teaches the concept of multiplication. I had a teacher tell me that her students had successfully memorized their facts but didn't know what they had acquired. After teaching them the skip-counting facts they understood what they had learned. I used to say that multiplying was fast adding. In reality it is fast adding of the same number. I can't multiply to find the solution to 1 + 4 + 6 + 9, but I can multiply to solve 4 + 4 + 4. Skip counting reinforces and teaches the concept of multiplication.

3. As a skill in itself, multiple counting is helpful. A pharmacist attending a workshop told me he skip counted when counting pills as they went into the bottles. Another man said he used the same skill for counting inventory at the end of every workday.

4. It teaches you the multiples of a number, which are so important when making equivalent fractions and finding common denominators. 2/5 = 4/10 = 6/15 = 8/20. The numbers 2 - 4 - 6 - 8 are the multiples of 2, and 5 - 10 - 15 - 20 are the multiples of 5.

5. In the curriculum, skip counting is reviewed in sequence form, asking students to fill in the blanks. __, __, __, 12, 15, __, __, __, 27, 30. This encourages them to find patterns in math, and patterns are key to understanding this logical and important subject.

One way to learn the skip counting facts is with the *Skip Count and Addition Songbook*. Included is a CD with the skip count songs from the twos to the nines sung to tunes taken from hymns and Christmas carols. Children enjoy it and it has proven very effective.

Another way to teach skip counting is by just plain counting, and then beginning to skip some of the numbers. This is where the concept originated. Look at example 1. Begin by counting each square: 1 - 2 - 3 - 4 - 5 - 6 . . . through 20. After this sequence is learned, skip the first number and just count the second number: 2 - 4 - 6 . . . through 20. This is skip counting. You know it as counting by twos.

When first introducing this method, you might try pointing to each square, and as you count the first number quietly ask the student(s) to say the second number loudly. Continue this practice, doing it more quietly each time, until you are just pointing to the first block silently while encouraging them to say the number loudly when you point to the second square. The student(s) see it, hear it, say it, and you can write the facts 2 - 4 - 6 . . . 20 as well.

Example 1

Skip count and write the number on the line. Say it out loud as you count and write. Then write the numbers in the spaces provided beneath the figure. By skipping the first space we come up with the expression "skip counting." The solution follows at the bottom of the page.

__2__ , __4__ , _____ , _____ , _____ , _____ , _____ , _____

Solution

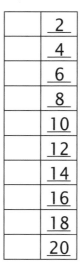

__2__ , __4__ , __6__ , __8__ , __10__ , __12__ , __14__ , __16__

LESSON 7

Addition with Regrouping (Carrying)

Don't forget place value—every value has its own place! To take it further, it's okay to visit another place, but there is no place like home! Let's add 34 and 28 to illustrate *regrouping* or "carrying." First add the units end to end and look for a ten. You can do this by putting a ten and a two beside the units, because 4 + 8 = 10 + 2 or 12. Leave the two units in the units column and "carry" the ten home. (Example 1 is continued on the next page.)

Mr. Ten can visit the units place, but he doesn't live there. I often interject, since I am 6' 5" myself, that Mr. Ten fell asleep on the couch, and since everything was only nine units long, his feet were hanging over the edge of the couch. So we carry him "home" to the tens place, picking up the ten bar and placing him with the other tens. We now have 6 tens and 2 units, or 62.

Example 1

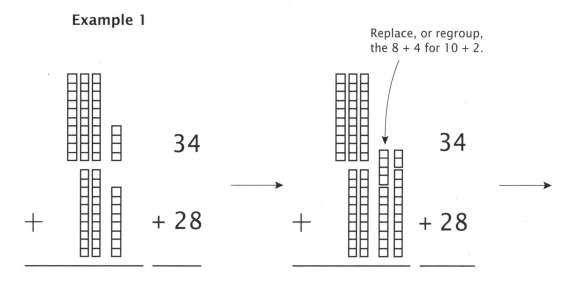

Replace, or regroup, the 8 + 4 for 10 + 2.

34
+ 28

34
+ 28

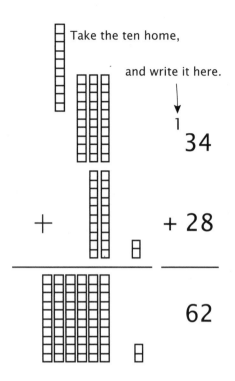

Take the ten home,

and write it here.

$$\begin{array}{r} 1 \\ 34 \\ +\ 28 \\ \hline 62 \end{array}$$

To reinforce the concept of place value, which is the key component in regrouping, I like to use place value notation to check the work. In the homework use both methods to make sure the student understands why we "carry" the one to the tens place.

$$\begin{array}{r} 1 \\ 34 \\ +\ 28 \\ \hline 62 \end{array} \longrightarrow \begin{array}{r} 10 \\ 30 + 4 \\ +\ 20 + 8 \\ \hline 60 + 2 \end{array}$$

Example 2

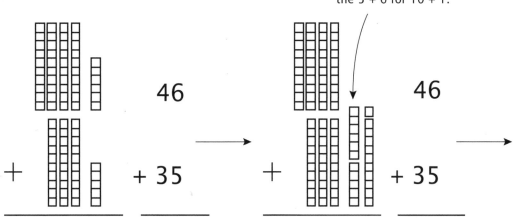

Replace, or regroup,
the 5 + 6 for 10 + 1.

46

+ 35

46

+ 35

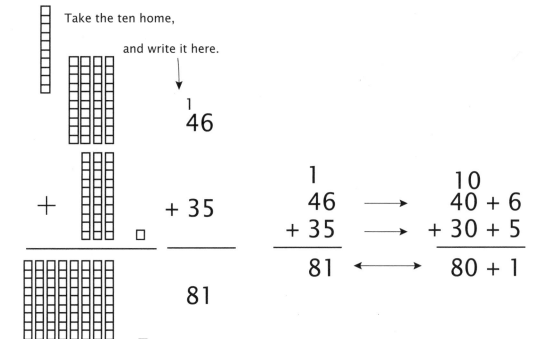

Take the ten home,

and write it here.

$$\begin{array}{r} 1 \\ 46 \\ + 35 \\ \hline 81 \end{array}$$

$$\begin{array}{r} 1 \\ 46 \\ + 35 \\ \hline 81 \end{array} \longrightarrow \begin{array}{r} 10 \\ 40 + 6 \\ + 30 + 5 \\ \hline 80 + 1 \end{array}$$

Example 3

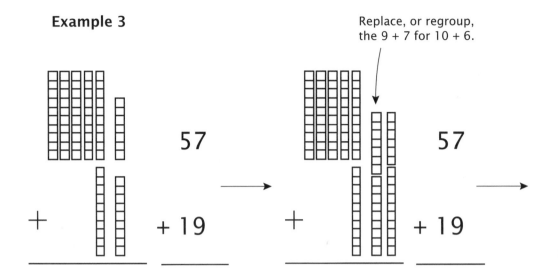

Replace, or regroup, the 9 + 7 for 10 + 6.

57

+ 19

57

+ 19

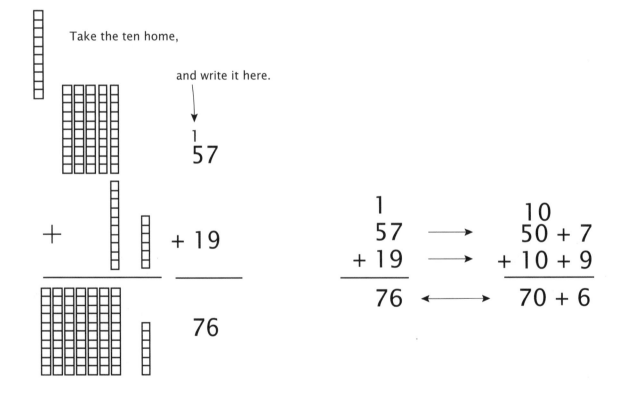

Take the ten home,

and write it here.

$$\begin{array}{r} 1 \\ 57 \\ + 19 \\ \hline 76 \end{array}$$

$$\begin{array}{r} 1 \\ 57 \\ + 19 \\ \hline 76 \end{array} \longrightarrow \begin{array}{r} 10 \\ 50 + 7 \\ + 10 + 9 \\ \hline 70 + 6 \end{array}$$

76

LESSON 8

Skip Count by 10
1 Penny = 1¢, 1 Dime = 10¢

After learning to skip count by 2, we need to learn skip counting by 10. Some practical examples are fingers on both hands, toes on both feet, and pennies in a dime. One of our objectives in this lesson is the concept that one dime has the same value as 10 pennies, and to apply this skill of counting by 10 to find out how many pennies are in several dimes. Notice that the multiples of ten (twenty, thirty, forty, etc.) all end in "ty." The suffix *ty* represents ten. So six-*ty* means six tens, and seven-*ty* means seven tens.

Example 1

As with the twos, skip count and write the number on the line. Say it out loud as you count and write. Then write the numbers in the spaces provided beneath the figure.

									10
									20
									30
									40
									50
									60

_____ , _____ , 30 , 40 , _____ , _____ , _____ , _____

Example 2

Fill in the missing information on the lines.

____ , ____ , ____ , 40 , ____ , ____ , 70 , 80 , 90 , ____

Solution

10 , 20 , 30 , 40 , 50 , 60 , 70 , 80 , 90 , 100

1 Penny = 1¢

Hold up a penny and teach that this is one penny or one cent. We write it as one penny or 1¢. The letter c with a line through it is the symbol used to represent cents. Two cents is 2¢. Nine cents is 9¢. With the blocks we can show one cent by holding up the green unit block.

1¢ = "one penny or one cent" =  =

Example 1

How many pennies are in 3¢?

three pennies or 3¢

1 Dime = 10¢

Another coin besides the penny is the dime. It is made of silver and is worth more than the copper penny. So one dime is equal in value to 10 pennies, even though it is smaller than a penny. With the blocks we represent the dime with the blue ten bar because one dime is 10 pennies, and one ten bar is the same as 10 green unit pieces.

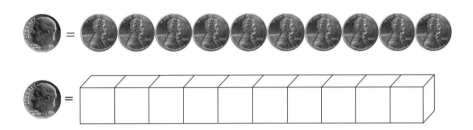

Example 2

How many pennies or cents are in three dimes?

 = "10 - 20 - 30" so 30 pennies or 30¢

Example 3

How many pennies or cents are in five dimes?

= "10 - 20 - 30 - 40 - 50"
so 50 pennies or 50¢

Skip Count by 5; 5¢ = 1 Nickel

In this lesson were teaching skip counting by five. Use the same techniques to introduce and teach this important skill that have worked before. Some practical examples are fingers on one hand, toes on a foot, pennies in a nickel, players on a basketball team, and sides of a pentagon. We can also remind the student that one nickel has the same value as five pennies, and apply the skill of counting by fives to find out how many pennies in several nickels.

Example 1

As with the twos, skip count and write the number on the line. Say it out loud as you count and write. Then write the numbers in the spaces provided beneath the figure.

				5
				10
				15
				20
				25
				30

____5____ , _____ , _____ , _____ , _____ , _____

Example 2

Fill in the missing information on the lines.

 5 , ___ , 15 , ___ , ___ , 30 , ___ , 40 , 45 , ___

Solution

 5 , 10 , 15 , 20 , 25 , 30 , 35 , 40 , 45 , 50

5¢ = One Nickel

The third coin to be learned is the nickel. It is made of several types of metal, looks like silver, and is worth more than the penny. One nickel is equal in value to five pennies and is a little larger than a penny. Using the blocks, we represent the nickel with the light blue five bar because one nickel is five cents, which is the same as five green unit pieces.

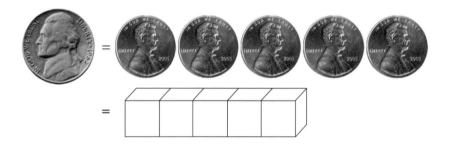

Example 3

How many pennies or cents are in four nickels?

= "5 - 10 - 15 - 20" so 20 pennies or 20¢

Example 4

How many pennies or cents are in seven nickels?

= "5 - 10 - 15 - 20 - 25 - 30 - 35"
so 35 pennies or 35¢

Example 5

How many pennies or cents are in three nickels?

= "5 - 10 - 15" so 15 pennies or 15¢

LESSON 10

Money: Decimal Point and Dollars

Money is a great way to reinforce the decimal system. It is the most effective illustration of all we have been learning since it uses the decimal system exactly, unlike imperial linear measurement (inches, feet, and yards), liquid measurement (pints, quarts, and gallons), or weight (ounces, pounds, and tons). It fits the manipulative blocks as well. The green units represent pennies, the ten bars represent 10 pennies or one dime, and the hundred squares represent 100 pennies or a dollar. To show $2.54, get two hundreds, five tens, and four units. Then teach that whereas the ¢ symbol indicates pennies or cents, the $ symbol represents dollars. The decimal point separates the dollars from the cents and is read "and" when reading from left to right. If you have a poster with Decimal Street (see lesson 1), put a black dot between the tens and hundreds for the decimal point. $2.54 is read as "two dollars and fifty-four cents." It may also be read as "Two dollars, five dimes, and four pennies." or as "two hundred fifty-four pennies." It is shown below.

Example 1

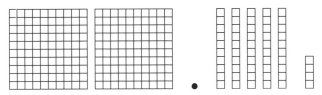

$ 2.54   "Two dollars and fifty-four cents"

Emphasize that 5 dimes is the same as 50 cents, so 5 dimes and 4 pennies is equal to 54 cents. When we read how many cents are to the right of the decimal point, we don't mention the dimes, only how many cents.

Example 2

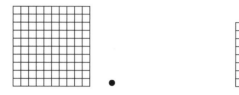

$ 1.08   "One dollar and eight cents"

Example 3

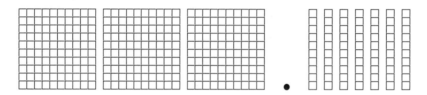

$ 3.70   "Three dollars and seventy cents"

Rounding to Hundreds
Multiple-Digit Addition with Regrouping

As we have been doing and always will do, add the units first. There may come a time when a student is so proficient at adding multiple digit numbers that he will add from the left, but this is down the road and not the way the decimal system operates. At this juncture, always add from right to left, from smaller to larger. Remember that you add units to units, tens to tens, and hundreds to hundreds. Whenever you add two numbers, always add the same values. "To combine, you must be the same kind." Here are two examples followed by the same problems worked out with place value notation and regular notation. If you have lined paper I suggest you turn it sideways to help keep the values in the proper places.

Rounding and Estimating to Hundreds

When adding large numbers, encourage the student to estimate the answer before solving it. We have learned how to round and estimate to the tens place, and now we want to increase our understanding by rounding and estimating to the hundreds place.

When you round a number to the nearest multiple of 100, there will be a number in the hundreds place but only a zero in the tens and units places, which are to the right of the hundreds place. It doesn't matter what numbers are present in the other place values, only the number to the immediate right of the place value being considered—in this case the tens place. This number determines whether to stay the same or be increased by one. I tell the students this is why we call it rounding, because the tens and units are going to be a "round" zero.

Example 1

Round 383 to the nearest hundreds place.

The first step is to find the two multiples of a hundred that are nearest to 383. The lower one is 300 and the higher one is 400; 383 is between 300 and 400. If the student has trouble finding these numbers, begin by placing your finger over the 83, so that all you have is a 3 in the hundreds place, which is 300. Then add one more to the hundreds to find the 400. I often write the numbers 300 and 400 above the number 383 on both sides as in figure 1.

Figure 1   300   400
                    383

Look at the number in the tens place. Does it fall in 0 through 4 or in 5 through 9? Since it is an 8, it is in the latter group, which means we round up to the next number, 400. Rounded to the nearest hundred, 383 is 400.

Example 2

Round 547 to the nearest hundreds place.

500   600   1. Find the multiples of one hundred nearest to 547.
   547

500   600   2. We know that 4 goes to the lower number, 500.
   547

In examples 3 and 4, the estimates are to the right in parentheses.

Example 3

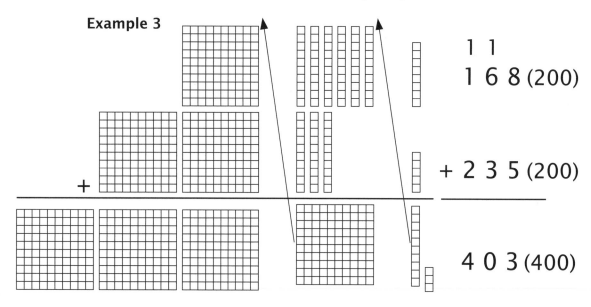

1 1
1 6 8 (200)

+ 2 3 5 (200)
_____

4 0 3 (400)

Example 3 (continued)   Five units plus 8 units equals 13, which is 1 ten and 3 units. We move the ten (or carry it) to the tens place as indicated by the arrow. Then 6 tens plus 3 tens plus the 1 ten from the result of adding in the units place equals 1 hundred. The 1 hundred is moved to the hundreds place as shown by the second arrow. Adding all the hundreds gives us the answer of 4 hundreds, 0 tens, and 3 units or 403. The picture below shows the result after regrouping.

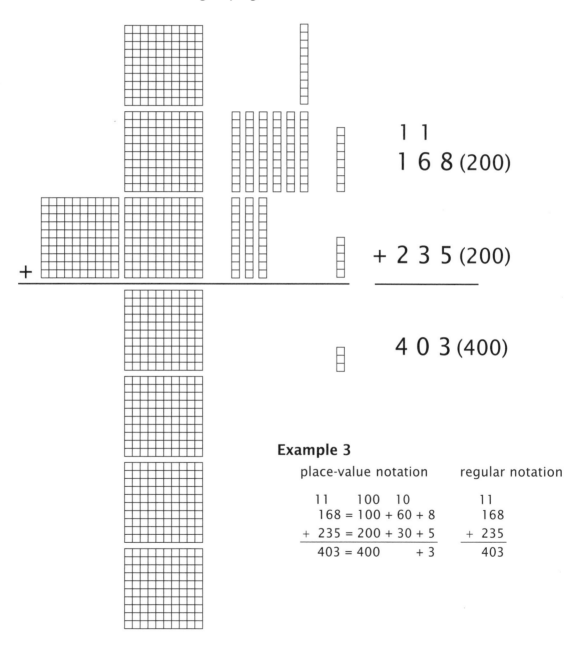

```
    1 1
    1 6 8 (200)

 +  2 3 5 (200)
    _____

    4 0 3 (400)
```

Example 3

place-value notation	regular notation

```
  11    100   10              11
  168 = 100 + 60 + 8          168
+ 235 = 200 + 30 + 5        + 235
  403 = 400       + 3         403
```

Example 4

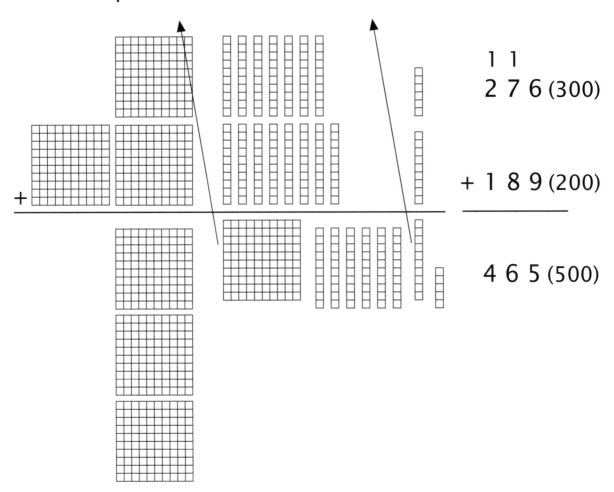

$$
\begin{array}{r}
\overset{1\ 1}{2\ 7\ 6}\ (300) \\
+\ 1\ 8\ 9\ (200) \\
\hline
4\ 6\ 5\ (500)
\end{array}
$$

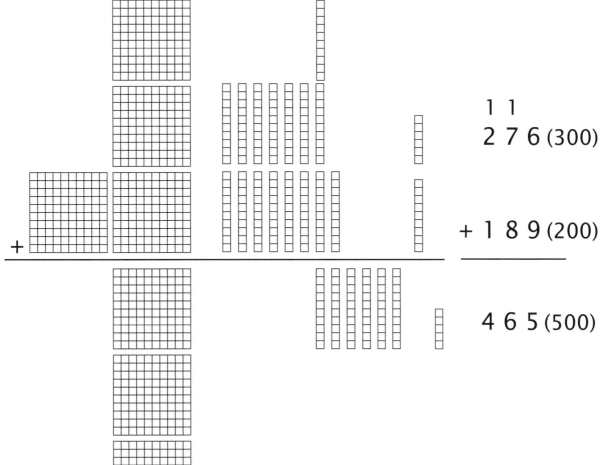

1 1
2 7 6 (300)

+ 1 8 9 (200)

4 6 5 (500)

Example 4

place-value notation   regular notation

```
   11    100   10              11
    276 = 200 + 70 + 6        276
  + 189 = 100 + 80 + 9      + 189
    465 = 400 + 60 + 5        465
```

Adding Money; Mental Math

The red hundreds square represents 100 pennies or one dollar; the blue ten bar shows us ten pennies or one dime: and the green unit cube, one penny. Adding money is the same as adding three-digit numbers. If we are adding pennies, the problem is identical. But we are adding dollars, dimes, and cents so we need a decimal point. We use the same blocks, but instead of regrouping 10 tens to form one hundred, we are talking about 10 dimes making one dollar. Ten pennies regroup to make one dime. In the examples, notice that we add the problem both ways, with dollars (and decimal points) and with pennies.

Emphasize the fact that to combine (add or subtract), things must be the same kind. So we add dollars to dollars, dimes to dimes, and units to units. That is why we line up the decimal points when adding, so all the place values, or money values, are lined up in the proper order.

Example 1 (continued on the next page)

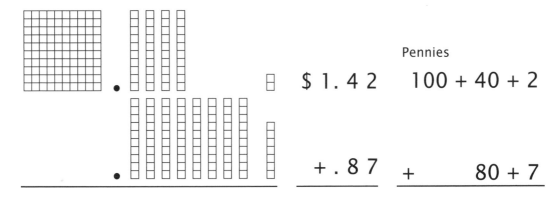

Solution

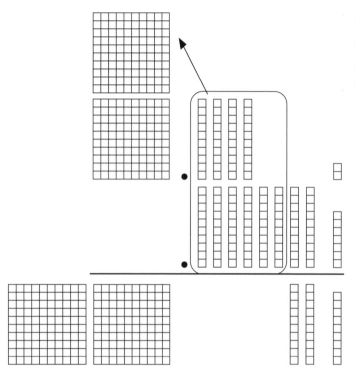

7¢ plus 2¢ equals 9¢. Eight dimes or 80¢, plus 4 dimes or 40¢, is the same as 1 dollar and 2 dimes or 120¢.

$$\begin{array}{r} 1 \\ \$1.42 \\ +\ .87 \\ \hline \$2.29 \end{array}$$

Pennies

$$\begin{array}{r} 100 \\ 100 + 40 + 2 \\ + \qquad 80 + 7 \\ \hline 200 + 20 + 9 \end{array}$$

Example 2

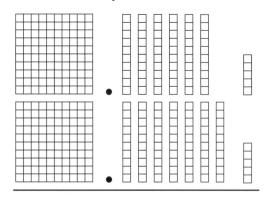

Pennies

$ 1 . 6 5   100 + 60 + 5

+

+ $ 1 . 7 5   + 100 + 70 + 5

Solution

5¢ plus 5¢ equals 10¢ which is 1 dime.

Six dimes or 60¢, plus 7 dimes or 70¢, plus the 1 dime from regrouping is the same as 1 dollar and 4 dimes or 140¢.

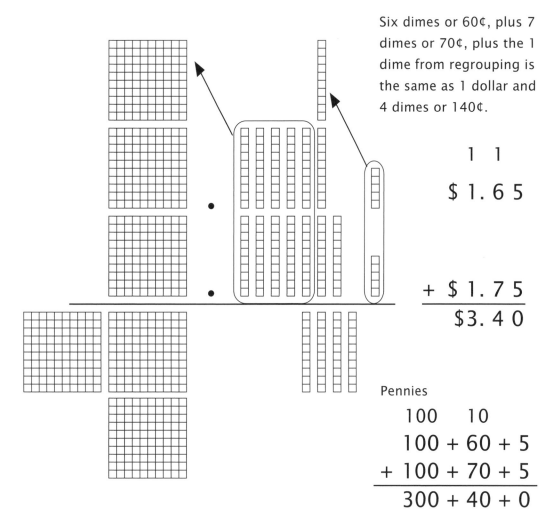

```
  1  1
$ 1 . 6 5
```

```
+ $ 1 . 7 5
  $ 3 . 4 0
```

Pennies

```
100    10
100 + 60 + 5
+ 100 + 70 + 5
300 + 40 + 0
```

Mental Math

These problems can be used to keep the facts alive in the memory and to develop mental math skills. The teacher should say the problem slowly enough so that the student comprehends, and then walk him through increasingly difficult exercises. The purpose is to stretch but not discourage. You decide where that line is!

Example 3

Two plus three plus one equals what number?

The student thinks, "Two plus three equals five, and five plus one equals six." Go slowly at first so the student can verbalize the intermediate step. As skills increase, he or she should be able to give just the answer.

Starting with this lesson, selected lessons in the instruction manual will have mental math problems for you to read aloud to your student. Look for them in lessons 18, 21, 24, and 27, and try a few at a time.

1. Seven plus two plus eight equals what numvber? (17)

2. One plus two plus four equals what number? (7)

3. Five plus three plus two equals what number? (10)

4. Six plus three plus six equals what number? (15)

5. Eight plus zero plus five equals what number? (13)

6. Three plus four plus one plus three equals what number? (11)

7. Two plus one plus four plus five equals what number? (12)

8. One plus two plus one plus two equals what number? (6)

9. Four plus one plus two plus nine equals what number? (16)

10. One plus five plus three plus three equals what number? (12)

Column Addition

Here is an example of column addition: $3 + 4 + 6 + 5 + 7 = 25$. The key is finding 10 in the list of numbers. In this case $3 + 7$ and $4 + 6$ make two tens. Adding the two tens and the five units equals 25.

Try this one: $1 + 2 + 3 + 4 + 5 + 6 + 7 + 8 + 9$. We have $\underline{1 + 9}$, $\underline{2 + 8}$, $\underline{3 + 7}$, $\underline{4 + 6}$, $+ 5$, which is four tens and five units, or 45. This is a good way to reinforce "carrying" or better still, "regrouping." Most of us were taught to carry "one" when, in reality, we are carrying "one ten" or taking ten home.

Example 1

```
  4      4
  7      7
  2      2 )10 )10
  8      8
  3      3
 +5    + 5
        29
```

There are 2 tens, which we carry to the tens place. They are equal to 20.

Add $4 + 5 = 9$ in the units place.

The answer is 29.

Example 2

```
      1 2
  26      26
  43      43    10
  31      31
  54      54    10
  17      17
 +25    +   25
          196
```

There are 2 tens, which we carry to the tens place. They are equal to 20.

Add 1 + 5 = 6 in the units place.

There are now 19 tens or 190, so we carry 1 hundred to the hundreds place and have 9 tens in the tens place.

The answer is 196.

Example 3

```
       3 1
  30         30
  24    100   24   10
  56    100   56
  74          74
  83    100   83
 +52    +    52
             319
```

There is 1 ten, which we carry to the tens place.

Add 2 + 3 + 4 = 9 in the units place.

There are now 31 tens or 310, so we carry 3 hundreds to the hundreds place, and have 1 ten in the tens place.

The answer is 319.

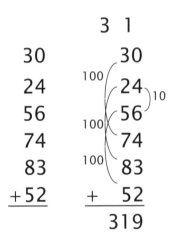

Measurement: 12 Inches = 1 Foot

The symbol for inches is " after the number. Eight inches is written as 8". The symbol for feet is ' after the number. Six feet is written as 6'. Give the student a 12" ruler. Or, the student may use a note card to make a ruler by placing it beside the ruler on this page and creating his or her own portable measuring instrument.

Each of the spaces between the vertical lines is one inch long. The whole line is five inches long. One foot is 12 inches long. When I don't have a ruler handy, I use my knuckle to estimate one inch. Have the student measure the distance between the two joints on each of his fingers to see which one is the closest to one inch.

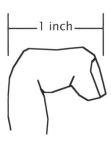

Example 1

Measure the length of the line and write your answer with the inch symbol.

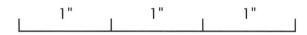

Solution

```
   1"        1"        1"
|____|____|____|
```

The line is 3" long.

Example 2

Using your ruler, note card, or knuckle, measure the length and width of this piece of paper.

Solution

It should be a little longer than 8 inches wide and 11 inches long, or 8" by 11".

Example 3

How many inches are in five feet?

Solution

Since there are 12" in 1', there are 12" + 12" + 12" + 12" + 12" in 5', and that is 60".

Perimeter

Often older students confuse *perimeter* and area because these terms have been taught simultaneously. One strategy to help them remember the difference is to see the word RIM in peRIMeter. Literally, it means the distance around. *Meter* means "distance" and *peri* means "around." I teach this concept by holding a rectangle in my hand and having the students imagine themselves to be ants crawling around the entire outside edge. Then I ask how many units around the rectangle?

Another strategy would be to take your student(s) outside and walk around the perimeter of a field or a yard. If weather is an issue consider walking around a room. This helps the student experience the concept of walking around rather than just hearing the information.

Perimeter is measured in linear units of measure. So the answer will be how many inches or feet. If the unit of measure is not given, let the answer be in "units."

Example 1   Find the perimeter of the rectangle.

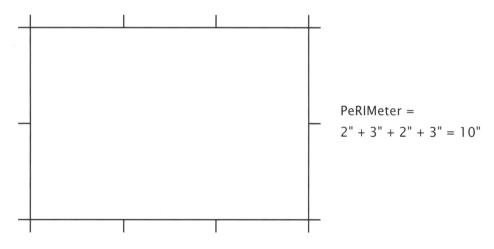

PeRIMeter =
2" + 3" + 2" + 3" = 10"

Example 2

Find the perimeter of the rectangle.

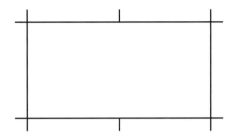

PeRIMeter = 2" + 1" + 2" + 1" = 6"

Example 3

Find the perimeter of the triangle.

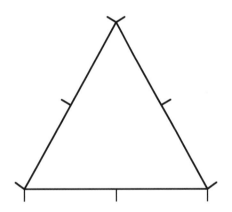

PeRIMeter = 2" + 2" + 2" = 6"

LESSON 16

Thousands and Place-Value Notation

A huge component in understanding multiple-digit addition is *place value*. The beginning value we call units. This is represented by the small green half-inch cube. The next largest place value is the tens place, shown with the blue ten bar. This is 10 times as large as a unit. The next value we come to as we move to the left is the hundreds, shown with the large red block. It is 10 times as large as a ten bar. Notice that as you move to the left, each value is 10 times as large as the preceding value. See figure 1.

Figure 1

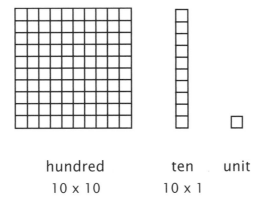

hundred	ten	unit
10 x 10	10 x 1	

When you name a number such as 247, the 2 tells you how many hundreds, the 4 indicates how many tens, and the 7 how many units. We read 247 as "two hundred four-ty (ty means ten) seven." The 2, 4, and 7 are digits that tell us *how many*. The hundreds, tens and units tell us *what kind*, or *what value*. Where the digit is written or what *place* it occupies tells us what *value* it is.

Notice that as the values progress from right to left they increase by a factor of 10. That is because we are operating in the base 10 system or the decimal system.

The next place value is the thousands place. It is 10 times 100. You can build a 1,000 by stacking 10 hundred squares and making a cube. You can also show a 1,000 by making a rectangle out of the cube that is 10 by 100 as in figure 2. You will see that I made it to a much smaller scale in order to show it. Also in this figure is 10,000. If you stick to rectangles, can you imagine what 100,000 would look like? It would be a rectangle 100 by 1,000. The factors are inside the rectangles.

Figure 2

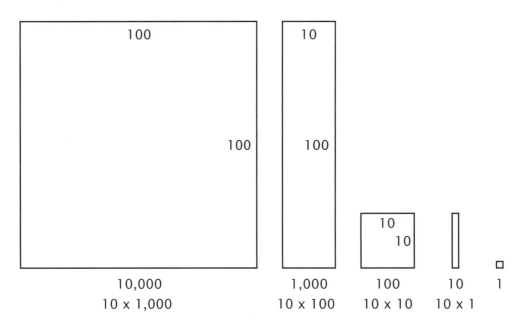

Notice that within the thousands there are 1, 10, and 100 just as in the units that are shown in figure 1. The commas separate the units and thousands. See figure 3.

Figure 3

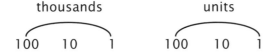

When saying these larger numbers, I like to think of the commas as having names. The first comma from the right is named "thousand." See example 1.

Example 1

Say 456,789.

"456 thousand 789" or
"four hundred fifty-six thousand, seven hundred eighty-nine"

Notice that you never say "and" when reading a large number. It is reserved for the decimal point in a succeeding book. Do you see that you never read a number larger than hundreds between the commas? This is because there are only three place between the commas. Practice saying and writing larger numbers.

Example 2

Say 13,762.

"13 thousand 762" or
"thirteen thousand, seven hundred sixty-two"

More Place-Value Notation

Place-value notation is a way of writing numbers that emphasizes the place value. The number in example 2 (13,762) looks like this in place value notation: 10,000 + 3,000 + 700 + 60 + 2. Each value is separated.

Note to Teacher

Place-value notation is used primarily to reinforce the role place value has in regrouping, whether carrying or borrowing. Once a student understands this relationship, the continued use of place-value notation may prove confusing. Even though it is included in the instruction manual, you don't have to continue using it with the student(s).

Example 3

Write 543,971 with place value notation.

500,000 + 40,000 + 3,000 + 900 + 70 + 1

Rounding to Thousands
Estimation

Rounding

Most of this lesson should be review since we have covered this material in previous lessons, but take your time to thoroughly digest what is new. Rounding is used in estimating as we add and subtract.

When you round a number to the nearest multiple of 1,000, there will be a number in the thousands place but only a zero in the places to the right of the thousands place. The number to the right of the place value we are rounding to determines whether to stay the same or increase by one. Perhaps this is why we call it rounding, because the hundreds, tens, and units are going to be a "round" zero.

Example 1

Round 4,299 to the nearest thousands place.

The first step is to find the two multiples of a thousand that are nearest to 4,299. The lower one is 4,000 and the higher one is 5,000. 4,299 is between 4,000 and 5,000.

Look at the number in the hundreds place. Does it fall in 0-4, or in 5-9? Since it is a 2, it is in the first group, which means the number in the thousands place stays the same and the other digits are "rounded" to zero. 4,299 rounded to the nearest thousand is 4,000.

4,000   4,299   5,000   1. Find the multiples of 1,000 nearest to 4,299.

(4,000)   4,299   5,000   2. We know that 2 stays the same, so the answer is 4,000.

Example 2

Round 6,502 to the nearest thousands place.

6,000   7,000
    6,502

1. Find the multiples of 1,000 nearest to 6,502.

6,000   ↗ (7,000)
    6,502

2. We know that 5 increases the thousands place by 1 to 7,000.

When rounding to thousands, remember to look only at the digit in the hundreds place to determine whether to stay the same or increase by one. The same rules apply to thousands that apply to hundreds and tens: if the digit in the hundreds place is a 0, 1, 2, 3, or 4, the number in the thousands place remains unchanged. If the digit is a 5, 6, 7, 8, or 9, then the digit in the thousands place increases by one.

When rounding to ten thousands, hundred thousands, or millions, consider only the number immediately to the right of the place to be rounded to in order to determine whether to stay the same or increase by one.

Example 3

Round 27,601 to the nearest ten thousands place.

20,000   (30,000)
   27,601

1. Find the multiples of 10,000 nearest to 27,601.

20,000   ↗ (30,000)
   27,601

2. The 7 to the right of 2 increases the ten thousands place by 1 to 30,000.

Estimation

The reason for rounding is to help us in estimating. In example 4 we'll use the results of examples 1 and 2 to solve an addition problem. The results of rounding are in the parentheses, as is the estimate. After adding the numbers, compare the answer to the estimate. It should be close, and we see that it is.

Example 4

$$
\begin{array}{ll}
4,299 & (4,000) \\
+\,6,502 & (7,000) \\ \hline
& (11,000)
\end{array}
\longrightarrow
\begin{array}{ll}
4,299 & (4,000) \\
+\,6,502 & (7,000) \\ \hline
10,801 & (11,000)
\end{array}
$$

In example 5 we'll use estimation to solve a subtraction problem. (This skill will be used in later lessons.) The results of rounding are in the parentheses, as is the estimate.

Example 5

Round 7,874 and 5,120 to the nearest thousands place.

7,000   7,874 ↗ (8,000)   (5,000) ↖ 5,120   6,000

$$
\begin{array}{ll}
7,874 & (8,000) \\
-\,5,120 & (5,000) \\ \hline
& (3,000)
\end{array}
\longrightarrow
\begin{array}{ll}
7,874 & (8,000) \\
-\,5,120 & (5,000) \\ \hline
2,754 & (3,000)
\end{array}
$$

Multiple-Digit Column Addition
Mental Math

Now we get a chance to use everything we have learned so far in this book. Place value, addition facts, regrouping, column addition, and making 10 all play a role in these BIG problems. Look for ways to make the big problems into small problems, and you can do this. Here are a few examples to help you. Try them yourself first, and then compare your work. Don't forget to estimate your answer first. This is shown in the parentheses.

The place value to the left of the hundreds is the thousands. One of the main problems that occurs is keeping the values in the proper places. If you are using lined paper, consider turning your paper sideways to make the best use of the lines for lining up place value.

Example 1

```
        2 2 2
 758     758      (800)
 342     342      (300)
 167     167      (200)
 532     532      (500)
+956    +956     (1000)
        ─────    (2,800)
        2,755
```

In the units place there is a 10 (8 + 2) plus 15, so you have 25. Put 2 tens with the other tens and leave 5 in the units place.

In the tens place there are 2 tens (5 + 5 and 6 + 4) which are really hundreds, and 5 more tens, or 250. Put 2 in the hundreds place and leave 5 in the tens place.

For the hundreds there are 2 tens (7 + 3 and 9 + 1) which are really 2 thousands, and 7 more hundreds (2,700), so put 2 in the thousands place and leave 7 in the hundreds place.

Example 2

```
        1 22
  263    263      (300)
   47     47      *(0)
  259    259      (300)
  558    558      (600)
+310   +310      (300)
       1,437    (1,500)
```

* rounding to hundreds

In the units place there is a 10 (or 3 + 7) plus 17, so you have 27. Put 2 tens with the other tens and leave 7 in the units place.

In the tens place there are 2 tens (5 + 5 and 6 + 4), which really are hundreds, and 3 more tens, or 230. Put 2 in the hundreds place and leave 3 in the tens place.

For the hundreds there is 1 ten (5 + 3 + 2), which is really a thousand, and 4 more hundreds, so put 1 in the thousands place and leave 4 in the hundreds place.

Example 3

```
        2 21
  527    527      (500)
   86     86      *(100)
  364    364      (400)
  411    411      (400)
+690   +690      (700)
       2,078    (2,100)
```

* rounding to hundreds

In the units place there is a 10 (or 6 + 4) plus 8, so you have 18. Put 1 ten with the other tens and leave 8 in the units place.

In the tens place there are 2 tens (8 + 2 and 9 + 1), which really are hundreds, and 7 more tens, or 270. Put 2 in the hundreds place and leave 7 in the tens place.

For the hundreds there are 2 tens (6 + 4 and 5 + 3 + 2), which are really 2 thousands, so put 2 in the thousands place and leave 0 in the hundreds place.

Mental Math

Here are some more questions to read to your student. These review subtraction. Remember to go slowly at first.

1. Nine minus one minus four equals what number? (4)

2. Eight minus six minus zero equals what number? (2)

3. Ten minus five minus two equals what number? (3)

4. Sixteen minus nine minus one equals what number? (6)

5. Fourteen minus five minus four equals what number? (5)

6. Six minus three minus one equals what number? (2)

7. Eleven minus seven minus three equals what number? (1)

8. Eighteen minus nine minus four equals what number? (5)

9. Fifteen minus one minus six equals what number? (8)

10. Ten minus two minus two equals what number? (6)

More Multiple-Digit Column Addition

The purpose of this lesson is to offer more practice, but you may not need it. The skills you know now are sufficient to do any addition problem. Take your time, put the values in the proper places, make 10, and you should be all set. After you do a few of these you will be thankful for calculators!

The place value to the left of the hundreds is the thousands, and to the left of that is the ten thousands. Rounding to the largest number in these problems will be thousands. So everything to the right of the thousands place will be zeros.

Example 1

```
         11 11
  8,534   8,534    (9,000)
  2,761   2,761    (3,000)
 +3,659  +3,659    (4,000)
         14,954   (16,000)
```

For the units there is a 10 (or 9 + 1) plus 4, so you have 14. Carry 1 ten and leave 4 in the units place.

In the tens place there is 1 ten (6 + 3 + 1) which is 1 hundred, and 5 more tens, or 150. Put 1 in the hundreds place and 5 in the tens place.

For the hundreds there are 19 hundreds, which is really 1 thousand, and 9 hundreds, so put 1 in the thousands place and leave 9 in the hundreds place.

In the thousands, there is 1 ten (8 + 2) which is 1 ten thousand, and 4 thousands, or 14,000. Place 1 in the ten thousands place and leave 4 in the thousands place.

Example 2

```
      21 1
 3,742     3,742    (4,000)
 9,555     9,555   (10,000)
+8,310    +8,310    (8,000)
          _____
          21,607   (22,000)
```

For the units there is 7.

In the tens place there is 1 ten (5 + 4 + 1) or 1 hundred. Put 1 in the hundreds place and 0 in the tens place.

For the hundreds there is 1 ten (7 + 3) plus 6, or 1 thousand and 6 hundred. Put 1 in the thousands place and leave 6 in the hundreds place.

In the thousands, there is 1 ten (9 + 1) and 11 more, or 2 ten thousands and 1 thousand, or 21,000. Place 2 in the ten thousands place and leave 1 in the thousands place.

Example 3

```
            11
 7,068     7,068    (7,000)
   460       460       (0)
+   37    +   37       (0)
          _____
           7,565    (7,000)
```

For the units there are 15. Carry 1 ten and leave 5 in the units place.

In the tens place there are 16 tens or 1 hundred, and 6 more tens, or 160. Put 1 in the hundreds place and 6 in the tens place.

For the hundreds there are 5.

In the thousands, there are 7.

Multiple-Digit Subtraction

When adding we have always said that to compare or combine you must have the same kind. Combining applies to adding and subtracting. When subtracting multiple-digit numbers, subtract from right to left. In the case of a multiple-digit number, subtract units from units, tens from tens, and hundreds from hundreds. In the examples, place-value notation is used as well as regular notation to emphasize this.

If using lined paper, I suggest turning it sideways to make the best use of the lines for differentiating the different place values.

Example 1

$$
\begin{array}{r}
957 \\
-342 \\
\hline
615
\end{array}
\qquad
\begin{array}{r}
900+50+7 \\
-300+40+2 \\
\hline
600+10+5
\end{array}
$$

Example 2

$$
\begin{array}{r}
598 \\
-208 \\
\hline
390
\end{array}
\qquad
\begin{array}{r}
500+90+8 \\
-200+00+8 \\
\hline
300+90+0
\end{array}
$$

Example 3

```
764    700+60+4
− 62  −     60+2
702    700+00+2
```

Example 4

```
 253    200+50+3
−120   −100+20+0
 133    100+30+3
```

When subtracting, you can easily check to see if your answer is correct by adding the subtrahend and the difference. (Remember that in a subtraction problem, the top number is the *minuend,* the second number is the *subtrahend*, and the answer is the *difference*.)

From Example 1

```
 957          957
−342         −342
 615          615
              957
```

To check this answer, I like to draw a curvy line under the 615 and then add 342 and 615. This sum should equal 957. I then draw an arrow to show that this is so.

From Example 2

```
  598 ←
 -208    )
  390    )
 ~~~~~
  598  )
```

Throughout the rest of this book, consider checking your answers this way. If the curvy line and arrow prove to be confusing, simply rewrite the problem beside the original.

From Example 1

```
  957        342
 -342      +615
  615        957
```

Telling Time: Minutes
Mental Math

When the student has mastered skip counting by five, he is ready to learn how to tell time. Teaching how to tell time with a clock that is not digital can be a challenge. We'll begin by taking 6 ten bars and explaining that there are 60 minutes in one hour. Next replace each ten bar with 2 five bars. If you have only the starter set of blocks, you can make this clock by using various combinations for five; for example, five units, or a two and a three, or a four and a unit. Take your 12 groups of five, and arrange them in a circle (which really is a dodecagon, or 12-sided polygon) so you have your 60 minutes in the shape of a clock. Beginning at the top, skip count by five going around the clock: 5 - 10 - 15 . . . 55 - 60.

Choose any long unit bar, turn it on its side so that it is smooth, and you have your minute hand. Point it at different areas, and beginning at the top, count your minutes.

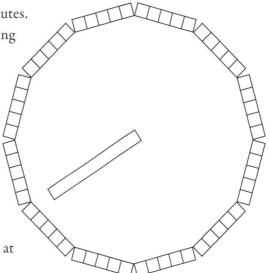

Even though we emphasize counting by five for reading the clock and finding the minutes, periodically remind the student that there are 60 possible minutes and they are not all multiples of five. So when giving problems on your own block clock, move the minute hand to numbers signifying 1, 2, 8, 34, or 51.

There is a template for your clock at the end of the student text.

To help the students see the progression of the minutes, build several partial clocks as in examples 1 and 2. Count the minutes by beginning at the top and moving around to the right, or clockwise.

Example 1
Count the minutes.
5 - 10, so 10 minutes

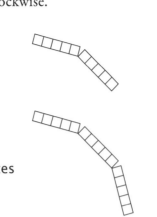

Example 2
Count the minutes.
5 - 10 - 15, so 15 minutes

Mental Math

Here are some more questions to read to your student. These combine addition and subtraction. Don't try the longer problems unless the student is comfortable with the shorter ones.

1. Four plus five, minus three equals what number? (6)

2. Seven minus two, plus six equals what number? (11)

3. Two plus nine, minus nine equals what number? (2)

4. Five minus three, plus ten equals what number? (12)

5. Ten minus eight, plus five equals what number? (7)

6. Seven minus six, plus eight, minus four equals what number? (5)

7. Four plus seven, minus three, minus seven equals what number? (1)

8. Eight plus eight, minus six, plus one equals what number? (11)

9. Ten minus two, plus four, minus six equals what number? (6)

10. Fourteen minus eight, plus two, minus four equals what number? (4)

Subtraction with Regrouping (Borrowing)

Now that we are comfortable adding and regrouping, we can apply this knowledge to the inverse: subtracting and regrouping. The following examples walk you through the process. The key is that when you don't have enough for subtraction to occur, you borrow from the next largest place value.

Example 1

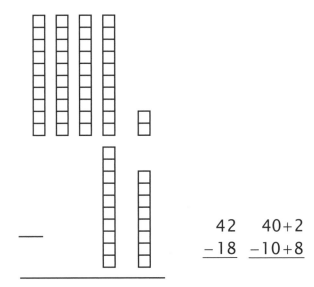

$$\begin{array}{r} 42 \\ -18 \\ \hline \end{array} \qquad \begin{array}{r} 40+2 \\ -10+8 \\ \hline \end{array}$$

The question is: "Can we count 8 from 2?" No! Two isn't big enough to have 8 counted from him. The number on top has to be the same or larger than the one underneath it. Ask the student if he has ever run to the neighbor's house to borrow a cup of flour, or milk, or whatever. Since there are not enough units, we need to go to our neighbor (the tens) and borrow one ten. (Continued on the next page.)

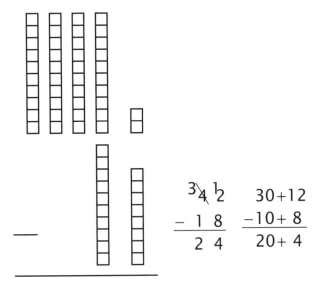

$$\begin{array}{r} \overset{3}{\cancel{4}}\,\overset{1}{2} \\ -\ 1\ 8 \\ \hline 2\ 4 \end{array} \qquad \begin{array}{r} 30+12 \\ -10+\ 8 \\ \hline 20+\ 4 \end{array}$$

Bringing the ten over to the units place, we now have 3 tens in the tens place. So we cross out the 4 and write a 3 in the tens place, and write a 1 beside the 2 in the units place; because with the ten we borrowed, we now have 12 in the units place.

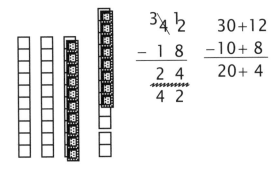

$$\begin{array}{r} \overset{3}{\cancel{4}}\,\overset{1}{2} \\ -\ 1\ 8 \\ \hline 2\ 4 \\ \hline 4\ 2 \end{array} \qquad \begin{array}{r} 30+12 \\ -10+\ 8 \\ \hline 20+\ 4 \end{array}$$

At this point we have enough to count up from 8 to 12. So we take the 8 unit bar and the one 10 bar and place them upside down on the 12 units and the 4 tens. We see that the difference is 2 tens and 4 units. Check by adding.

Example 2

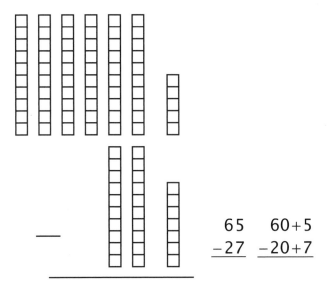

$$65 \quad 60+5$$
$$-27 \quad -20+7$$

The question is, "Can we count 7 from 5?" No! Five isn't big enough to have 7 counted from him. The number on top has to be the same or larger than the one underneath it.

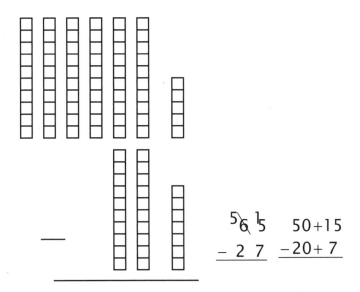

$$\overset{5}{\cancel{6}}\,\overset{1}{5} \quad 50+15$$
$$-\ 2\ 7 \quad -20+\ 7$$

Bringing the ten over to the units place, we now have 5 tens in the tens place. So we cross out the 5 and write a 4 in the tens place and a 1 beside the 5 in the units place. We do this because with the ten we borrowed, we now have 15 in the units place.

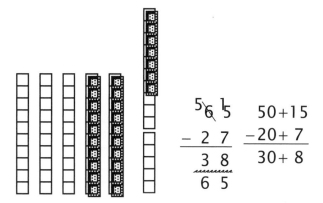

$$\begin{array}{r} {}^{5}\cancel{6}\,{}^{1}5 \\ -\ 2\ 7 \\ \hline 3\ 8 \\ \wedge\!\wedge\!\wedge\!\wedge\!\wedge \\ 6\ 5 \end{array}\qquad \begin{array}{r} 50+15 \\ -20+\ 7 \\ \hline 30+\ 8 \end{array}$$

At this point we have enough to count up from 7 to 15. So we take the 5 unit bar and the two 10 bars and place them upside down on the 15 units and the 5 tens. We see that the difference is 3 tens and 8 units. Check by adding.

LESSON 23

Telling Time: Hours

When the minutes are mastered, we can add the hours by placing a green unit bar at the end of the first five bar (outside the circle), pointing away from the center of the clock. To distinguish between the minutes and hours, I leave the minute unit bars right side up, but the hours I place upside down with the hollow side showing. You can still see the color and how many hours there are, but distinguish the minutes and the hours. Place your orange bar (upside down so the hollow side is showing) at the end of the second five bar. Continue this process with all the unit bars through 12. There is an illustration on the next page.

Choose a smaller unit bar than your minute hand for an hour hand. Turn it upside down so the student makes the connection between the hour hand and the hours, since both are upside down with the hollow side showing.

Now position the hour hand so that it points between the 2 and the 3. This is the critical point for telling time. Is it 2 or is it 3? I've explained this to many children, with success, by aiming the hour hand at the 2 and saying, "He just had his second birthday and is now 2." Then I move the hand towards the 3 a little, and ask, "How old is he now? Is he still 2?" "Yes." Then I move it a little further and ask if he's still 2. "Yes." I do this until it is almost pointing to 3 and ask the question, "What about the day before his next birthday; how old is he?" "Still 2." He is almost 3 but still 2. Practice this skill until the student can confidently identify which hour it is by moving the hour hand around the clock.

When the hour hand is mastered, put the hours and minutes together. Have the student first identify the hour, then skip count to find the minutes. Use the practice sheets with both hands shown. Finally, when this is mastered, have them tell time without the manipulatives by looking at a clock face.

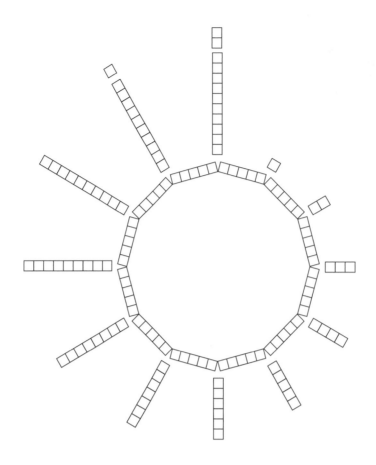

There is a removable clock template at the end of the student text for you to use if you wish.

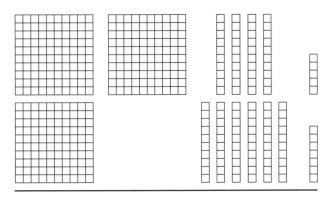

Subtraction: Three-Digit Numbers
Mental Math

When borrowing from the hundreds to the tens, move a hundred over from the hundreds place, cross out the digit in that place, and replace it with a digit that is one less. Then change the hundred in the tens place to ten 10 bars. Put a one beside the digit in the tens place to show that it has been increased by 10. Continue by borrowing, or regrouping, one ten to make 10 units. (Example 1 is continued on the next page.)

Example 1

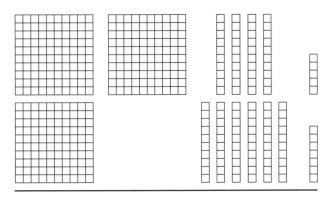

$$\begin{array}{r} 245 \\ -167 \\ \hline \end{array} \qquad \begin{array}{r} 200+40+5 \\ -100+60+7 \\ \hline \end{array}$$

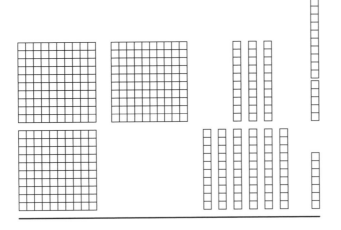

Can we count 7 from 5? No. So we take 1 ten and add it to the 5 units to make 15 units. There are only 3 tens now in the top number, the minuend.

$$2\,^3\!\!\!\not{4}\,^1\!\!\!\not{5} \qquad 200+30+15$$
$$-1\ 6\ 7 \qquad -100+60+\ 7$$

We can't subtract 6 tens from 3 tens. We take 1 hundred and add it to the 3 tens to make 130 or 13 tens. There is only 1 hundred now in the top number, the minuend. After all this, we subtract! The addition check is also shown.

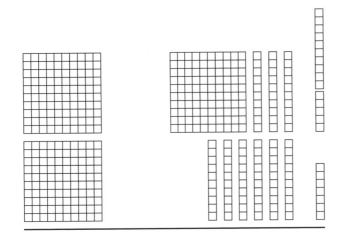

$$^1\!\!\not{2}\ ^{13}\!\!\not{4}\ ^1\!\!\not{5} \qquad 100+130+15$$
$$-\ 1\ \ 6\ \ 7 \qquad -100+\ 60\ +\ 7$$
$$\underline{7\ \ 8} \qquad \underline{70\ +\ 8}$$
$$2\ \ 4\ \ 5$$

This system of crossing out one from a larger place and making the next smaller place 10 more will always work. In the decimal system, when moving between the place values from right to left, or smaller to larger, you always increase by a factor of 10. So 10 units make one ten, and 10 tens make one hundred. When moving in the other direction, one hundred makes 10 tens and one ten makes 10 units.

Example 2 (continued on the next page)

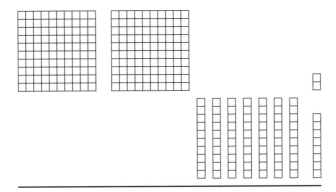

$$
\begin{array}{r}
202 \\
-\ 78
\end{array}
\qquad
\begin{array}{r}
200+\ 0+2 \\
-\quad\ 70+8
\end{array}
$$

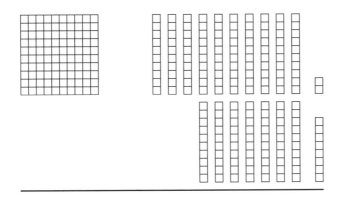

We can't subtract 8 units from 2 units, but there are no tens to regroup and make units. So we take 1 hundred and regroup to make 10 tens.

$$
\begin{array}{r}
{}^{1}\cancel{2}\,{}^{1}02 \\
-\ \ 78
\end{array}
\qquad
\begin{array}{r}
100+100+2 \\
-\quad\ 70+8
\end{array}
$$

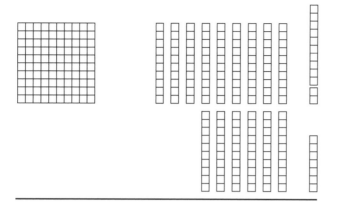

Then we take 1 ten and add it to the 2 units to make 12 units. After this is all done, we subtract!

$$
\begin{array}{r}
9 \\
\overset{1}{\cancel{2}}\,\overset{1}{\cancel{0}}\,2 \\
-\;\;\;7\;\;8 \\
\hline
1\;\;2\;\;4 \\
2\;\;0\;\;2
\end{array}
$$

$$
\begin{array}{r}
100+90+12 \\
-\qquad 70+8 \\
\hline
100+20+\;4
\end{array}
$$

Mental Math

These problems combine addition and subtraction.

1. Two plus three, minus two, plus six equals what number? (9)

2. Five plus two, minus one, plus seven equals what number? (13)

3. Nine minus five, plus four, plus six equals what number? (14)

4. Four plus ten, minus seven, plus five equals what number? (12)

5. Fifteen minus eight, minus two, plus four equals what number? (9)

6. Two plus ten, minus nine, plus ten equals what number? (13)

7. Seven plus seven, minus seven, minus seven equals what number? (0)

8. Eight plus six, plus one, minus nine equals what number? (6)

9. Seventeen minus eight, minus four, plus zero equals what number? (5)

10. Seven plus three, minus eight, plus four equals what number? (6)

Ordinal Numbers; Tally Marks
Days and Months

In our math so far we have been using cardinal numbers, which tell how many. *Ordinal numbers* are used to tell what order. Notice the "ord" in ordinal and order. Other than first, second, and third, ordinal numbers usually end in *th*. I like to use this opportunity to teach days of the week and months of the year. The first day of the week is Sunday, the first month of the year is January, etc. If your student doesn't know the days and months, take some time to learn them before moving to the next lesson. Each day, ask the student to give the date and tell, using ordinal numbers, which month it is and which day. Example: Christmas is in the twelfth month and on the twenty-fifth day.

Choose a birthday of any student or member of the family, a famous person, or important date in history and ask the same questions. If my birthday is on the eighth month and the first day, what is the date? The date is August 1.

first	Sunday	January	31
second	Monday	February	28 or 29
third	Tuesday	March	31
fourth	Wednesday	April	30
fifth	Thursday	May	31
sixth	Friday	June	30
seventh	Saturday	July	31
eighth		August	31
ninth		September	30
tenth		October	31
eleventh		November	30
twelfth		December	31

Days in a Month

My grandfather taught me a neat way to tell how many days are in a month without using the poem "30 days has September . . ." Holding up your two fists, and beginning on the left, recite the months. The knuckles, or mountains, are 31 days, and the valleys between the knuckles are 30 days (except February, which is 28 days, or 29 days in a leap year). Notice that when you put your fists together, July and August are both mountains. There is no valley between them, so they each have 31 days.

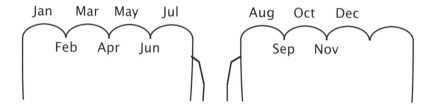

Example 1

How many days are in September?

Counting from the left, we end up in a valley, so that is 30 days.

Example 2

What is the tenth month?

Beginning with January as the first month, we end up with October.

Example 3

What is the fourth day of the week?

Beginning with Sunday as the first day of the week, we end up with Wednesday.

10 Commandments

Besides the days of the week and months in a year, some teachers use this opportunity to teach the Ten Commandments. The fifth commandment is to honor your mother and your father.

Tally Marks

When you use tally marks, you can only count to four using one line for one, two lines for two, three lines for three, and four lines for four. To show five, you put a slash diagonally through four lines. Tally marks are very useful when keeping track of slowly changing information. Since cowboys used a tally book to keep track of their cattle, I think of cows slowly moving through a gate and a cowboy sitting on a fence making one mark for each cow passing beneath him. Or perhaps you are on a trip, and you want to count all the red cars you pass. Each time you pass a red car you make a line until you get to the fifth one, and then you make a slash.

Look at the chart below to see how to represent the numbers 1 to 20. Then study the examples as we change from tally marks to a number and then from a number to tally marks.

#	tally	#	tally
1	⎮	11	卌 卌 ⎮
2	⎮⎮	12	卌 卌 ⎮⎮
3	⎮⎮⎮	13	卌 卌 ⎮⎮⎮
4	⎮⎮⎮⎮	14	卌 卌 ⎮⎮⎮⎮
5	卌	15	卌 卌 卌
6	卌 ⎮	16	卌 卌 卌 ⎮
7	卌 ⎮⎮	17	卌 卌 卌 ⎮⎮
8	卌 ⎮⎮⎮	18	卌 卌 卌 ⎮⎮⎮
9	卌 ⎮⎮⎮⎮	19	卌 卌 卌 ⎮⎮⎮⎮
10	卌 卌	20	卌 卌 卌 卌

Example 1

Change the number 7 to tally marks.

$7 = 5 + 2 =$ 𝍷𝍷𝍷𝍷̸ | |

Example 2

Change 𝍷𝍷𝍷𝍷̸ 𝍷𝍷𝍷𝍷̸ | | | | to a number.

𝍷𝍷𝍷𝍷̸ 𝍷𝍷𝍷𝍷̸ | | | | $= 5 + 5 + 4 = 14$

Subtraction: Four-Digit Numbers

When we borrow from the thousands to the hundreds, we cross out the digit in the thousands place and replace it with a digit that is one less. Next, we change the thousand in the hundreds place to 10 hundreds, then put a one beside the digit in the hundreds place to show that it has been increased by 10. This is what we have always done when regrouping tens or hundreds. It shows the uniformity of the base 10 system.

Example 1
Solve 4,581 − 1,397.

    7
4,5 8̸ 1̸   4,000+500+70+11
−1,3 9 7 −1,000+300+90+ 7

We can't subtract 7 units from 1 unit. We take 1 ten and regroup to make 10 units.

   4 17
4,5̸ 8̸ 1̸   4,000+400+170+11
−1,3 9 7 −1,000+300+ 90+ 7

We can't subtract 9 tens from 7 tens. We take 1 hundred and regroup to make 10 more tens.

   4 17
4,5̸ 8̸ 1̸   4,000+400+170+11
−1,3 9 7 −1,000+300+ 90+ 7
 3,1 8 4   3,000+100+ 80+ 4
 4,5 8 1

The hundreds and thousands are okay, so we can subtract. We could have subtracted the units after they were changed in step 1. Either way will work.

Example 2

Solve 3,062 - 1,549.

```
        5
    3,0 6̷ 2      3,000+000+50+12
   −1,5 4 9      −1,000+500+40+ 9
  ────────      ──────────────────
```

We can't take 9 units from 2 units. We take 1 ten, regroup, and make 12 units.

```
    2   5
    3̷,0̷ 6̷ 2      2,000+1,000+50+12
   −1,5 4 9      − 1,000+ 500+40+ 9
  ────────      ──────────────────
```

We can subtract 4 tens from 5 tens so that is fine. But we can't take 5 hundreds from 0 hundreds.
We regroup 1 thousand to make 10 hundreds.

```
    2   5
    3̷,0̷ 6̷ 2      2,000+1,000+50+12
   −1,5 4 9      − 1,000+ 500+40+ 9
  ────────      ──────────────────
    1,5 1 3      1,000+ 500+10+ 3
    3,0 6 2
```

The thousands are okay, so we can subtract.

Subtraction: Money
Mental Math

We know that the red hundreds square represents one dollar, the blue tens bar represents one dime, and the green units cube represents one penny. Subtracting money is the same as subtracting three-digit numbers except for the decimal point. We use the same blocks, but instead of regrouping a hundred to make 10 tens, we regroup one dollar to make 10 dimes. In the same vein, one dime is regrouped to form 10 pennies.

Example 1 (continued on the next page)

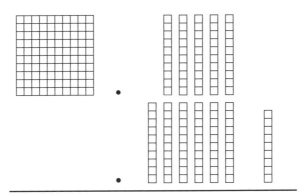

$$\begin{array}{r} \$1.50 \\ -\ \ .69 \\ \hline \end{array} \qquad \begin{array}{r} 100+50+0 \\ -\ \ \ \ \ 60+9 \\ \hline \end{array}$$

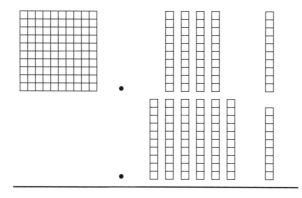

We can't subtract 9¢
from 0¢. We take 1
dime and add it to 0¢
to make 10¢.

    4
$1.5 ⁰
− .6 9

100+40+10
−   60+ 9

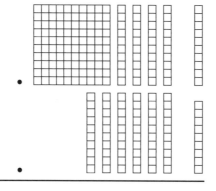

We can't subtract 6
dimes from 4 dimes,
or 60¢ from 40¢. We
take 1 dollar and add
it to 40¢ to make 140¢
or 14 dimes.

   14
$ 1.5 ⁰
− .6 9
$. 8 1

140+10
− 60+ 9
 80+ 1

Mental Math

Here are some more questions to read to your student. These combine addition and subtraction. You may shorten these if your student is not yet ready for the longer questions.

1. Four plus four, minus six, plus nine equals what number? (11)

2. Eight plus two, minus three, plus three equals what number? (10)

3. Seven minus five, plus six, plus ten equals what number? (18)

4. Six plus eight minus nine, plus five, equals what number? (10)

5. Seventeen minus ten, minus four, plus four equals what number? (7)

6. Nine plus eight, minus nine, plus one equals what number? (9)

7. Five plus nine, minus six, minus five equals what number? (3)

8. Ten plus two, minus seven, plus nine equals what number? (14)

9. Fifteen minus one, minus five, plus eight equals what number? (17)

10. Five plus nine, minus four, plus eight equals what number? (18)

Subtraction: Multiple-Digit Numbers

The only new part of this lesson is subtracting five digits instead of four. As we have already learned, when we need to regroup, we simply cross out the number in the ten thousands place and decrease it by one. Then put a one beside the number in the thousands place and increase it by 10. Everything else is the same as what you have already learned.

Example 1
Solve 34,085 - 19,760.

```
    3
  3 4,085        30,000+3,000+1,000+80+5
 −19,760        −10,000+9,000+ 700 +60+0
```

The units and tens are okay. We can't subtract 7 hundreds from 0 hundreds, so we take 1 thousand and regroup to make 10 hundreds.

```
  3 4,085        20,000+13,000+1,000+80+5
 − 1 9,760      −10,000+ 9,000 + 700 +60+0
   1 4,3 2 5     10,000+ 4,000 + 300 +20+5
   3 4,0 8 5
```

We can't subtract 9 thousand from 3 thousand, so we take 1 ten thousand and regroup to make 10 thousands. Now we subtract and check.

Example 2

Solve 97,518 - 69,227.

```
      4
 97,5̷¹18        90,000+7,000+400+110+8
-69,2 27       -60,000+9,000+200+ 20+7
_____       _____
```

The units are okay. We can't take 2 tens from 1 ten, so we take 1 hundred and regroup to make 10 more tens. Subtract 7 hundreds from 0 hundreds.

```
 8    4
 8̷ ¹7,5̷¹18       80,000+17,000+400+110+8
- 6 9,2 27      -60,000+ 9,000+200+ 20+7
_____      _____
  2 8,2 91       20,000+ 8,000+200+ 90+1
^^^^^^^^^^^
  9 7,5 18
```

We can't subtract 9 thousand from 7 thousand, so we take 1 ten thousand and regroup to make 10 more thousands. Now we subtract and check.

Reading Gauges and Thermometers

Reading Gauges

Many measuring instruments do not give you all the information and require you to figure out some of it. Here are some gauges found around my home and in my car. When doing these problems, first figure out how much is between each line. Then start at 0 or E and count how many spaces to the arrow. Be careful that you don't count the lines. Instead count how many spaces are between the lines. For each example, there are two problems and the solutions. Read where the arrow is pointing and figure out the answer.

Example 1

On the 20-gallon gas gauge in my car, how much gas is left on each gauge?

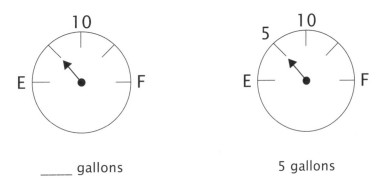

_____ gallons   5 gallons

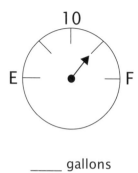

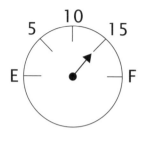

_____ gallons

_15_ gallons

Example 2

For the coal stove in the basement, how hot is it on each gauge?

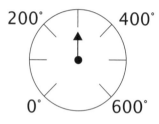

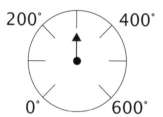

_____ °

_300_ °

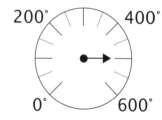

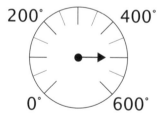

_____ °

_500_ °

Example 3

Reading my speedometer, how fast am I going on each gauge?

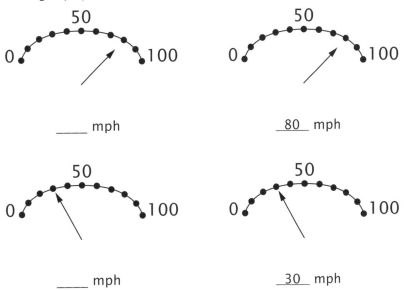

_____ mph   _80_ mph

_____ mph   _30_ mph

Thermometers

The key to reading scales on a thermometer or graph is finding out what each space stands for on the scale. Skip counting, as we have been learning it, comes to our aid as we find out what the pattern is. In example 4 we recognize skip counting by twos: 2, 4, 6, __, __, 12, __, 16. Filling in the missing numbers (0, 2, 4, 6, 8, 10, 12, 14, 16), we discover the temperature is 10°. The floating circle above and to the right of 10 denotes degrees.

Example 4

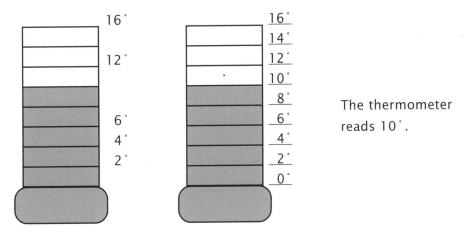

The thermometer reads 10°.

Example 5

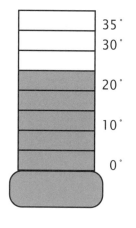 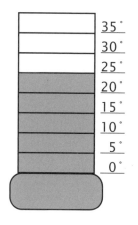

The thermometer reads 25°.

Example 6

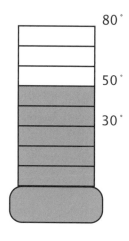 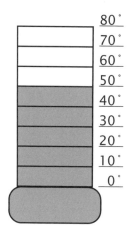

The thermometer reads 50°.

Bar Graphs and Line Graphs

Bar Graphs

When doing a bar graph, use your bars to represent the number values. To make this lesson real and practical, choose from the following list of ideas or make up some of your own. Keep track of money earned in a week, rainfall, how fast you grow, temperature, home runs of your favorite baseball team per month, or the number correct on your homework assignments for your graph. Make it REAL.

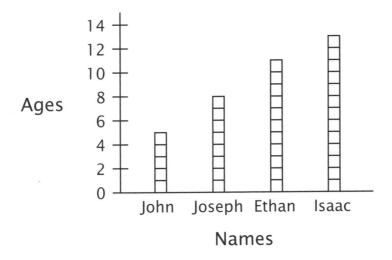

If I were to graph the ages of my boys, it would look like this. Your objects, or data, are on the bottom and labeled. Your scale is on the left. For larger numbers you can let each block or unit represent 5, 10, or 100.

Line Graphs

To draw a line graph, do the same procedure as in a bar graph except draw a dot at the top of each bar, and then remove the bar. This leaves just dots that you connect to form a line graph.

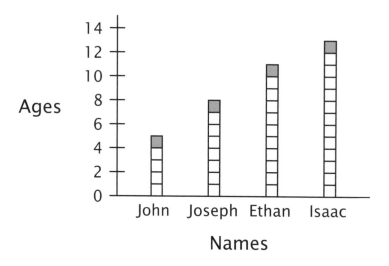

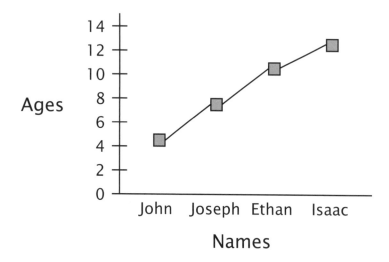

Nineovers or Striking Nines

There is no check for adding large numbers or adding columns of numbers. While the skill presented here is not foolproof, it is a quick and effective way to test the answer to an addition problem. Until the advent of calculators, it was widely known and used by accountants and bookkeepers.

There are various names for this method of checking; some call it casting out nines, and others refer to it as striking nines. I have coined the term *nineovers*, since it is based on dividing by nine and counting what is left over (the remainder). Since the student hasn't learned how to divide by nine, the explanation is for the teacher and can be shared with the student at the appropriate time. We'll begin with the concept, or the why, of nineovers.

If you divided 27 by 9, the answer would be 3 with a remainder of 0. But we could have predicted this remainder by adding up the digits of 27: 2 + 7 = 9. We will learn in multiplication by nine that a characteristic shared by all the multiples of nine is that the digits add up to nine or a multiple of nine. Notice this in 18, 27, 36, 45, 54, and so on. The digits add up to nine. In 963, the digits add up to 18, which is a multiple of nine. If you were to divide 963 by 9, you would have an answer of 107 with a remainder of 0.

Consider the number 274. If you divide it by 9, you get 30 with a remainder of 4. It is 2 + 7 = 9, with 4 left over. You can predict the remainders, or "nineovers," of a number divided by nine by adding up the digits and subtracting nine. In the number 138, the nineover is 3 because 1 + 8 = 9. If you were adding 274 + 138, the nineover for the sum should be 7 (4 from 274 and 3 from 138). If it were 6 or 8, then you would expect an error in your calculations and go back to rework the problem. See figure 1 on the next page.

Figure 1

$$
\begin{array}{r l}
274 & \rightarrow 4 \\
\underline{+138} & \rightarrow 3 \\
412 & \rightarrow 7
\end{array}
$$

Example 1

$$
\begin{array}{r l l}
274 \rightarrow & \cancel{2} + \cancel{7} + 4 & \rightarrow 4 \\
\underline{+138} \rightarrow & \cancel{1} + 3 + \cancel{8} & \rightarrow 3 \\
412 \rightarrow & 4 + 1 + 2 & \rightarrow 7
\end{array}
$$

Strike out the combinations of 9 in each addend. Then add the nineovers: 4 + 3 = 7.

Compare this answer with adding up the digits in the sum: 4 + 1 + 2 = 7. They agree.

$$
\begin{array}{r l l r}
274 \rightarrow & 2 + 7 + 4 = 13 & \rightarrow & 1 + 3 = 4 \\
\underline{+138} \rightarrow & 1 + 3 + 8 = 12 & \rightarrow & 1 + 2 = 3 \\
412 \rightarrow & 4 + 1 + 2 = 7 & \rightarrow & 7
\end{array}
$$

Add the digits in the addends until they are a single digit. 2 + 7 + 4 = 13 and then 1 + 3 = 4.

Compare this answer with adding up the digits in the sum: 4 + 1 + 2 = 7. They agree.

Study the following examples and practice with this skill until you feel comfortable with it. Then consider the third technique for applying this method of checking given in example 7.

Example 3

$$2731 \rightarrow \cancel{2} + \cancel{7} + 3 + 1 \rightarrow 4$$
$$4251 \rightarrow \cancel{4} + 2 + \cancel{5} + 1 \rightarrow 3$$
$$\underline{+2120} \rightarrow 2 + 1 + 2 + 0 \rightarrow 5$$
$$9{,}102 \qquad \cancel{9} + 1 + 0 + 2$$

Strike out the nines in the addends. As you add the nineovers 4 + 3 + 5, strike the 4 + 5 so the nineover is 3.

Compare this answer with adding up the digits in the sum: 1 + 2 = 3. They agree.

Example 4

$$2731 \rightarrow 2+7+3+1 \rightarrow 13 \quad 1+3 = 4$$
$$4251 \rightarrow 4+2+5+1 \rightarrow 12 \quad 1+2 = 3$$
$$\underline{+2120} \rightarrow 2+1+2+0 \rightarrow \quad 5 \qquad = 5$$
$$9{,}102 \qquad 9+1+0+2 \rightarrow 12 \quad 1+2 = 3$$

Add the digits in the addends until they are a single digit. 4 + 3 + 5 = 12 and then 1 + 2 = 3.

Compare this answer with adding up the digits in the sum: 1 + 2 = 3. They agree.

Example 5

$$727 \rightarrow \quad \cancel{7} + \cancel{2} + 7 \rightarrow 7$$
$$342 \rightarrow \quad \cancel{3} + \cancel{4} + \cancel{2} \rightarrow 0$$
$$\underline{+\ \ 565} \rightarrow \quad 5+6+5 \rightarrow 16 \quad 1+6 = 7$$
$$1{,}625 \qquad \cancel{1} + \cancel{6} + \cancel{2} + 5$$

Strike out the nines in the addends. Add the digits in the addends until they are a single digit. 7 + 7 = 14 and 1 + 4 = 5.

Compare this answer with adding up the digits in the sum: 5. They agree.

Example 6

$$727 \rightarrow 7+2+7 \rightarrow 16 \quad 1+6=7$$
$$342 \rightarrow 3+4+2 \rightarrow 9 \quad = 0$$
$$\underline{+\quad 565} \rightarrow 5+6+5 \rightarrow 16 \quad 1+6=7$$
$$1{,}625 \quad 1+6+2+5 \quad 14 \quad 1+4=5$$

Add the digits in the addends until they are a single digit. $7 + 7 = 14$ and $1 + 4 = 5$.

Compare this answer with adding up the digits in the sum: 5. They agree.

Example 7

$$727 \rightarrow (7+2)+7$$
$$342 \rightarrow 3+4+2$$
$$\underline{+\quad 565} \rightarrow 5+6+5 \rightarrow 5$$
$$1{,}625 \quad (1+6+2)+5 \rightarrow 5$$

In this example you can strike nines anywhere among the addends. This agrees with the remainder in the sum: $5 = 5$.

Student Solutions

Lesson Practice 1A

1. 146
   "one hundred forty-six"
2. 254
   "two hundred fifty-four"
3. 2 tens and 8 units
   "twenty-eight"
4. 3 hundreds, 3 tens, and 6 units
   "three hundred thirty-six"
5. $3+1=4$
6. $5+0=5$
7. $2+8=10$
8. $0+7=7$
9. $5+2=7$
10. $2+1=3$
11. $1+5=6$
12. $6+2=8$
13. $4+2=6$
14. $2+9=11$
15. $7+2=9$

Lesson Practice 1B

1. 139
   "one hundred thirty-nine"
2. 325
   "three hundred twenty-five"
3. 3 tens and 1 unit
   "thirty-one"
4. 2 hundreds, 2 tens, and 8 units
   "two hundred twenty-eight"
5. $9+3=12$
6. $8+7=15$
7. $9+5=14$
8. $8+3=11$
9. $5+8=13$
10. $6+9=15$
11. $9+4=13$
12. $8+9=17$

13. $9+7=16$
14. $6+8=14$
15. $8+4=12$

Lesson Practice 1C

1. 193
   "one hundred ninety-three"
2. 262
   "two hundred sixty-two"
3. 4 tens and 9 units
   "forty-nine"
4. 2 hundreds, 8 tens, and 3 units
   "two hundred eighty-three"
5. $8+8=16$
6. $7+7=14$
7. $7+6=13$
8. $4+4=8$
9. $5+5=10$
10. $6+5=11$
11. $3+3=6$
12. $2+2=4$
13. $6+6=12$
14. $3+4=7$
15. $9+9=18$

Lesson Practice 1D

1. 105
   "one hundred five"
2. 246
   "two hundred forty-six"
3. 7 tens and 2 units
   "seventy-two"
4. 4 hundreds, 1 ten, and 7 units
   "four hundred seventeen"
5. $8+1=9$
6. $7+3=10$
7. $3+6=9$

8. $1 + 9 = 10$
9. $6 + 4 = 10$
10. $5 + 5 = 10$
11. $7 + 2 = 9$
12. $5 + 4 = 9$
13. $2 + 8 = 10$
14. $3 + 7 = 10$
15. $4 + 6 = 10$

Systematic Review 1E

1. 318
   "three hundred eighteen"
2. 273
   "two hundred seventy-three"
3. 5 tens
   "fifty"
4. 1 hundred and 9 tens
   "one hundred ninety"
5. $4 + 7 = 11$
6. $7 + 5 = 12$
7. $3 + 5 = 8$
8. $5 + 7 = 12$
9. $5 + 3 = 8$
10. $7 + 4 = 11$
11. $9 + 8 = 17$
12. $5 + 3 = 8$
13. $7 + 6 = 13$
14. $8 + 2 = 10$
15. $1 + 0 = 1$

Systematic Review 1F

1. 130
   "one hundred thirty"
2. 45
   "forty-five"
3. 3 hundreds and 6 units
   "three hundred six"
4. 2 hundreds, 2 tens, and 2 units
   "two hundred twenty-two"

5. $1 + 1 = 2$
6. $7 + 2 = 9$
7. $9 + 5 = 14$
8. $3 + 8 = 11$
9. $6 + 6 = 12$
10. $7 + 8 = 15$
11. $5 + 4 = 9$
12. $3 + 5 = 8$
13. $5 + 8 = 13$
14. $9 + 4 = 13$
15. $3 + 3 = 6$
16. $6 + 7 = 13$
17. $7 + 5 = 12$
18. $4 + 8 = 12$

Lesson Practice 2A

1. 4, 12, 20
2. 6, 31, 206
3. 162, 55, 4
4. 136, 16, 3
5. 8, 9, 10, 11, 12
6. 10, 9, 8, 7, 6
7. 71, 72, 73, 74, 75

Lesson Practice 2B

1. 5, 8, 16
2. 25, 63, 100
3. 7, 17 107
4. 11, 89, 200
5. 30, 10, 6
6. 80, 20, 10
7. 245, 61, 9
8. 100, 84, 3
9. 6, 5, 4, 3, 2
10. 22, 23, 24, 25, 26
11. 64, 65, 66, 67, 68
12. 17, 16, 15, 14, 13

Lesson Practice 2C

1. 6, 50, 100
2. 18, 32, 95
3. 10, 40, 400
4. 16, 65, 243
5. 125, 15, 5
6. 170, 106, 9
7. 99, 48, 22
8. 120, 114, 89
9. 29, 28, 27, 26, 25
10. 43, 44, 45, 46, 47
11. 30, 31, 32, 33, 34
12. 9, 8, 7, 6, 5

Systematic Review 2D

1. 5, 15, 17
2. 8, 11, 40
3. 75, 43, 12
4. 200, 150, 110
5. 79, 80, 81, 82, 83
6. 4 tens and 7 units
    "forty-seven"
7. 1 hundred, and 9 units
    "one hundred nine"
8. $9+9=18$
9. $7+8=15$
10. $4+5=9$
11. $2+3=5$
12. $4+8=12$
13. $5+7=12$
14. $3+8=11$
15. 121 is larger than 112, so Sam
16. $4+1=5$ sandwiches

Systematic Review 2E

1. 5, 18, 63
2. 100, 200, 400
3. 56, 44, 14
4. 200, 105, 50

5. 9, 8, 7, 6, 5
6. 9 tens and 8 units
    "ninety-eight"
7. 2 hundreds, 7 tens, and 6 units
    "two hundred seventy-six"
8. $6+6=12$
9. $5+9=14$
10. $2+7=9$
11. $4+6=10$
12. $8+9=17$
13. $5+6=11$
14. $1+7=8$
15. 7 is smaller than 16, so daisies
16. $7+9=16$ vehicles

Systematic Review 2F

1. 19, 39, 99
2. 10, 60, 80
3. 299, 74, 18
4. 180, 48, 21
5. 88, 89, 90, 91, 92
6. 2 hundreds and 4 tens
    "two hundred forty"
7. 1 ten and 6 units
    "sixteen"
8. $7+7=14$
9. $2+5=7$
10. $3+7=10$
11. $0+4=4$
12. $8+8=16$
13. $6+7=13$
14. $4+9=13$
15. 62 is larger than 26, so story books
16. $6+8=14$ coins

Lesson Practice 3A

1. $4>3$
2. $7=7$
3. $6>4$
4. $11<21$

5. $14 = 14$
6. $10 > 9$
7. $63 > 36$
8. $12 = 12$
9. $6 < 7$
10. $12 > 10$
11. $50¢ < 90¢$
    $90¢ > 50¢$
12. Denise: $5 + 4 = 9$
    Michael: $5 + 5 = 10$
    $9 < 10$, so Michael did more.
13. $12 > 8$, so Ethan read fewer.
14. $14 = 14$

Lesson Practice 3B

1. $5 < 7$
2. $2 < 4$
3. $10 > 9$
4. $13 > 10$
5. $18 > 17$
6. $11 = 11$
7. $27 < 72$
8. $14 < 15$
9. $6 = 6$
10. $12 = 12$
11. Ray: $2 + 1 = 3$
    Brother: $3 + 2 = 5$
    $5 > 3$, so brother had more.
12. $9 > 8$, so Alexa got more right.
13. $10 = 10$
14. $150 > 105$

Lesson Practice 3C

1. $9 < 14$
2. $5 < 6$
3. $9 > 7$
4. $25 > 15$
5. $16 = 16$
6. $15 < 17$
7. $31 > 13$

8. $12 = 12$
9. $7 > 5$
10. $11 < 12$
11. $10 < 110$; Rachel has fewer.
12. $65 > 55$; more on clothes
13. $10 = 10$
14. $213 < 231$

Systematic Review 3D

1. $13 > 12$
2. $57 < 75$
3. 7, 8, 13
4. 50, 70, 90
5. 28, 29, 30, 31, 32
6. 2 hundreds, 1 ten, and 1 unit
   "two hundred eleven"
7. 3 tens and 6 units
   "thirty-six"
8. $2 + 9 = 11$
9. $5 + 5 = 10$
10. $3 + 6 = 9$
11. $2 + 8 = 10$
12. $6 + 9 = 15$
13. $3 + 4 = 7$
14. $0 + 6 = 6$
15. $452 > 254$
16. $5 + 3 = 8$ children

Systematic Review 3E

1. $8 = 8$
2. $91 > 19$
3. 200, 12, 2
4. 105, 51, 15
5. 17, 18, 19, 20, 21
6. 1 hundred and 4 units
   "one hundred four"
7. 1 ten and 3 units
   "thirteen"
8. $4 + 4 = 8$
9. $1 + 6 = 7$

10. $2+4=6$
11. $0+1=1$
12. $9+9=18$
13. $7+8=15$
14. $4+7=11$
15. $3 > 2$, so more brothers
16. $3+3=6$ miles

Systematic Review 3F

1. $6 < 7$
2. $15 = 15$
3. 300, 63, 3
4. 197, 91, 74
5. 13, 12, 11, 10, 9
6. 1 hundred, 3 tens, and 5 units
   "one hundred thirty-five"
7. 2 hundreds and 4 tens
   "two hundred forty"
8. $7+9=16$
9. $2+2=4$
10. $4+9=13$
11. $3+7=10$
12. $1+1=2$
13. $5+8=13$
14. $7+7=14$
15. $13=13$
16. $6+8=14$ pages

Lesson Practice 4A

1. 40
2. 40
3. done
4. 60
5. done
6.   (10)
   $\underline{+(10)}$
     (20)

7.   (30)
   $\underline{+(20)}$
     (50)
8.   (50)
   $\underline{+(40)}$
     (90)
9.

3	5	8
6	1	7
9	6	15

10. $(20)+(30)=(50)$ books

Lesson Practice 4B

1. 40
2. 20
3. 60
4. 90
5.   (20)
   $\underline{+(20)}$
     (40)
6.   (50)
   $\underline{+(30)}$
     (80)
7.   (30)
   $\underline{+(20)}$
     (50)
8.   (60)
   $\underline{+(20)}$
     (80)
9.

3	2	5
5	2	7
8	4	12

10. $(60)+(30)=(90)$ dollars

Lesson Practice 4C

1. 30
2. 80
3. 70
4. 60

5. (30)
    +(30)
    (60)

6. (20)
    +(10)
    (30)

7. (40)
    +(50)
    (90)

8. (20)
    +(40)
    (60)

9.
4	5	9
3	1	4
7	6	13

10. (30) + (30) = (60) in

Systematic Review 4D
1. 80
2. 50
3. (50)
    +(20)
    (70)
4. (20)
    +(30)
    (50)
5.
6	1	7
2	3	5
8	4	12

6. 4 = 4
7. 16 < 21
8. 4 + 6 = 10
9. 1 + 9 = 10
10. 6 + 9 = 15
11. 5 + 7 = 12
12. (20) + (10) = (30) hrs
13. 9 + 8 = 17 battles
14. 2 + 2 = 4
    4 + 6 = 10 birds

Systematic Review 4E
1. 50
2. 60
3. (40)
    +(10)
    (50)
4. (70)
    +(10)
    (80)
5.
2	1	3
7	0	7
9	1	10

6. 14 > 13
7. 99 < 105
8. 3 + 9 = 12
9. 1 + 8 = 9
10. 5 + 5 = 10
11. 2 + 7 = 9
12. 3 + 3 = 6 cones
13. (20) + (30) = (50) worms
14. 279 < 410

Systematic Review 4F
1. 90
2. 80
3. (50)
    +(30)
    (80)
4. (50)
    +(20)
    (70)
5.
3	3	6
2	1	3
5	4	9

6. 18 > 16
7. 198 < 201
8. 9 + 5 = 14
9. 2 + 6 = 8
10. 4 + 3 = 7
11. 3 + 8 = 11
12. 9 + 2 = 11 cards

13. $(50)+(40)=(90)$ radios
14. $45 < 405$

Lesson Practice 5A

1. done
2. $40+7=47$
3. $100+40+3$
4. $60+5$
5. done
6. $70+9=79$
7. $40+5=45$
8. $200+60+9=269$
9. $24+33=57$
10. $16+12=28$ dollars

Lesson Practice 5B

1. $100+20+4=124$
2. $50+9=59$
3. $300+60+5$
4. $40+1$
5. $70+9=79$
6. $20+6=26$
7. $60+8=68$
8. $300+50+5=355$
9. $133+10=143$ dollars
10. $23+23=46$ blackbirds

Lesson Practice 5C

1. $300+30+5=355$
2. $60+8=68$
3. $100+70+2$
4. $20+7$
5. $40+4=44$
6. $90+6=96$
7. $90+9=99$
8. $400+40+8=448$
9. $12+36=48$ students
10. $41¢+46¢=87¢$

Systematic Review 5D

1. $500+30+1$
2. $10+8$
3. $40+3=43$
4. $400+20+2=422$
5. done
6. $(40)+(20)=(60)$
   $43+21=64$
7. $7<12$
8. $71>17$
9. $6+\underline{6}=12$
10. $7+\underline{4}=11$
11. $5+\underline{2}=7$
12. $120+254=374$ birds

Systematic Review 5E

1. $100+10+4$
2. $30+9$
3. $30+9=39$
4. $300+50+3=353$
5. $(70)+(10)=(80)$
   $68+11=79$
6. $(50)+(30)=(80)$
   $51+34=85$
7.

5	6	11
7	1	8
12	7	19

8. $15=15$
9. $103<331$
10. $7+\underline{3}=10$
11. $9+\underline{7}=16$
12. $5+\underline{4}=9$
13. $(20)+(20)=(40)$ questions
14. $212+362=574$ miles

Systematic Review 5F

1. $300+10+4$
2. $70+2$
3. $90+9=99$

4. $400+90+9=499$
5. $(70)+(20)=(90)$
   $72+15=87$
6. $(60)+(20)=(80)$
   $61+17=78$
7.

2	8	10
6	3	9
8	11	19

8. $6=6$
9. $86>68$
10. $8+\underline{5}=13$
11. $4+\underline{1}=5$
12. $9+\underline{2}=11$
13. $665+133=798$ animals
14. $42>24$

Lesson Practice 6A
1. 2, 4, 6, 8, 10, 12, 14, 16, 18, 20
2. 2, 4, 6, 8, 10, 12, 14, 16, 18, 20
3. 2, 4, 6, 8, 10, 12, 14, 16, 18, 20
4. 2, 4, 6, 8, <u>10</u> cookies
5. 2, 4, 6, 8, 10, <u>12</u> X's
6. 2, 4, 6, 8, 10, 12, 14, 16, <u>18</u> shoes

Lesson Practice 6B
1. 2, 4, 6, 8, 10, 12, 14, 16, 18, 20
2. 2, 4, 6, 8, 10, 12, 14, 16, 18, 20
3. 2, 4, 6, 8, 10, 12, 14, 16, 18, 20
4. 2, 4, <u>6</u> cookies
5. 2, 4, 6, 8, 10, 12, <u>14</u> X's
6. 2, 4, 6, 8, <u>10</u> ears

Lesson Practice 6C
1. 2, 4, 6, 8, 10, 12, 14, 16, 18, 20
2. 2, 4, 6, 8, 10, 12, 14, 16, 18, 20
3. 2, 4, 6, 8, 10, 12, 14, 16, 18, 20
4. 2, 4, 6, 8, 10, 12, 14, <u>16</u> snowflakes

5. 2, 4, 6, 8, <u>10</u> arms
6. 2, 4, <u>6</u> books

Systematic Review 6D
1. 2, 4, 6, 8, 10, 12, 14, 16, 18, 20
2. $20+2=22$
3. $200+80+9=289$
4. $(10)+(20)=(30)$
   $13+15=28$
5. $(30)+(40)=(70)$
   $32+41=73$
6. $8>6$
7. $108<801$
8. $5+\underline{3}=8$
9. $8+\underline{7}=15$
10. $9+\underline{4}=13$
11. 2, 4, 6, 8, 10, 12, <u>14</u> times
12. $7+2=9$
    $9+3=12$ times

Systematic Review 6E
1. 2, 4, 6, 8, 10, 12, 14, 16, 18, 20
2. $30+8=38$
3. $600+90+9=699$
4. $(50)+(20)=(70)$
   $48+21=69$
5. $(50)+(10)=(60)$
   $54+12=66$
6. $10>8$
7. $295<592$
8. $3+\underline{0}=3$
9. $5+\underline{3}=8$
10. $4+\underline{5}=9$
11. $126+132=258$ dollars
12. 2, 4, 6, 8, 10, <u>12</u> fawns

Systematic Review 6F

1. 2, 4, 6, 8, 10, 12, 14, 16, 18, 20
2. $60 + 3 = 63$
3. $600 + 90 + 6 = 696$
4. $(10) + (30) = (40)$
   $11 + 33 = 44$
5. $(50) + (30) = (80)$
   $53 + 25 = 78$
6. $7 = 7$
7. $42 > 24$
8. $7 + \underline{2} = 9$
9. $6 + \underline{6} = 12$
10. $7 + \underline{7} = 14$
11. 2, 4, 6, $\underline{8}$ inches
12. $11 + 6 = 17$ pennies

9.
```
  1
  75
+ 16
  91
```

10. $59 + 37 = 96$ marbles

Lesson Practice 7B

1.
```
  1      10
  35     30 + 5
+ 25    + 20 + 5
  60     60 + 0
```

2.
```
  1      10
  59     50 + 9
+ 24    + 20 + 4
  83     80 + 3
```

3.
```
  1      10
  45     40 + 5
+ 28    + 20 + 8
  73     70 + 3
```

4.
```
  1      10
  76     70 + 6
+ 15    + 10 + 5
  91     90 + 1
```

5.
```
  1      10
  24     20 + 4
+  9    + 00 + 9
  33     30 + 3
```

6.
```
  1      10
  25     20 + 5
+ 39    + 30 + 9
  64     60 + 4
```

7.
```
  1
  24
+ 48
  72
```

8.
```
  1
  67
+  8
  75
```

9.
```
  1
  56
+ 24
  80
```

10. $26 + 38 = 64$ cookies

Lesson Practice 7A

1. done

2.
```
  1      10
  46     40 + 6
+ 35    + 30 + 5
  81     80 + 1
```

3.
```
  1      10
  73     70 + 3
+ 18    + 10 + 8
  91     90 + 1
```

4.
```
  1      10
  35     30 + 5
+ 37    + 30 + 7
  72     70 + 2
```

5.
```
  1      10
  26     20 + 6
+ 55    + 50 + 5
  81     80 + 1
```

6.
```
  1      10
  38     30 + 8
+ 44    + 40 + 4
  82     80 + 2
```

7.
```
  1
  47
+ 25
  72
```

8.
```
  66
+ 33
  99
```

Lesson Practice 7C

1.
$$
\begin{array}{cc}
1 & 10 \\
25 & 20+5 \\
+65 & +60+5 \\
\hline
90 & 90+0
\end{array}
$$

2.
$$
\begin{array}{cc}
34 & 30+4 \\
+45 & +40+5 \\
\hline
79 & 70+9
\end{array}
$$

3.
$$
\begin{array}{cc}
1 & 10 \\
72 & 70+2 \\
+18 & +10+8 \\
\hline
90 & 90+0
\end{array}
$$

4.
$$
\begin{array}{cc}
1 & 10 \\
68 & 60+8 \\
+\;8 & +00+8 \\
\hline
76 & 70+6
\end{array}
$$

5.
$$
\begin{array}{cc}
1 & 10 \\
34 & 30+4 \\
+46 & +40+6 \\
\hline
80 & 80+0
\end{array}
$$

6.
$$
\begin{array}{cc}
1 & 10 \\
36 & 30+6 \\
+45 & +40+5 \\
\hline
81 & 80+1
\end{array}
$$

7.
$$
\begin{array}{c}
1 \\
48 \\
+\;7 \\
\hline
55
\end{array}
$$

8.
$$
\begin{array}{c}
1 \\
24 \\
+66 \\
\hline
90
\end{array}
$$

9.
$$
\begin{array}{c}
1 \\
18 \\
+53 \\
\hline
71
\end{array}
$$

10. $16+8 = 24$ cards

Systematic Review 7D

1.
$$
\begin{array}{cc}
1 & 10 \\
49 & 40+9 \\
+45 & +40+5 \\
\hline
94 & 90+4
\end{array}
$$

2.
$$
\begin{array}{cc}
1 & 10 \\
36 & 30+6 \\
+36 & +30+6 \\
\hline
72 & 70+2
\end{array}
$$

3.
$$
\begin{array}{cc}
1 & 10 \\
68 & 60+8 \\
+25 & +20+5 \\
\hline
93 & 90+3
\end{array}
$$

4.
$$
\begin{array}{c}
55 \\
+14 \\
\hline
69
\end{array}
$$

5.
$$
\begin{array}{c}
1 \\
77 \\
+\;7 \\
\hline
84
\end{array}
$$

6.
$$
\begin{array}{c}
95 \\
+\;3 \\
\hline
98
\end{array}
$$

7. 2, 4, 6, 8, 10, 12, 14, 16, 18, 20

8. $12 < 13$

9. $113 > 103$

10. $8 + \underline{6} = 14$

11. $6 + \underline{4} = 10$

12. $9 + \underline{6} = 15$

13.

1	4	5
7	3	10
8	7	15

14. 2, 4, 6, 8, 10, $\underline{12}$ jokes

15. $14 + 17 = 31$ trips

Systematic Review 7E

1.
$$
\begin{array}{c}
12 \\
+13 \\
\hline
25
\end{array}
$$

2.
$$
\begin{array}{c}
1 \\
37 \\
+28 \\
\hline
65
\end{array}
$$

3.
$$
\begin{array}{c}
63 \\
+36 \\
\hline
99
\end{array}
$$

4. $22 + 48 = 70$

5. $82 + 9 = 91$

6. $52 + 6 = 58$

7. 2, 4, 6, 8, 10, 12, 14, 16, 18, 20

8. 20

9. 20

10. $9 + \underline{5} = 14$

11. $5 + \underline{1} = 6$

12. $7 + \underline{6} = 13$

13.

2	6	8
5	8	13
7	14	21

14. $11 > 8$

15. $125 + 172 = 297$ dollars

Systematic Review 7F

1.
$$\begin{array}{r} 43 \\ +26 \\ \hline 69 \end{array}$$

2.
$$\begin{array}{r} 1 \\ 19 \\ +18 \\ \hline 37 \end{array}$$

3.
$$\begin{array}{r} 62 \\ +32 \\ \hline 94 \end{array}$$

4. $15 + 36 = 51$

5. $29 + 8 = 37$

6. $79 + 6 = 85$

7. 2, 4, 6, 8, 10, 12, 14, 16, 18, 20

8. 50

9. 70

10. $7 + \underline{5} = 12$

11. $8 + \underline{3} = 11$

12. $9 + \underline{8} = 17$

13.

7	8	15
6	5	11
13	13	26

14. 2, 4, 6, 8, 10, 12, 14, <u>16</u> bows

15. $(30) + (10) = (40)$
$28 + 13 = 41$ animals

Lesson Practice 8A

1. 10, 20, 30, 40, 50, 60, 70, 80, 90, 100
2. 10, 20, 30, 40, 50, 60, 70, 80, 90, 100
3. 10, 20, 30, 40, 50, 60, <u>70¢</u>
4. 10, 20, 30, <u>40¢</u>
5. 10, 20, 30, 40, <u>50¢</u>
6. 10, 20, 30; 3 packages

Lesson Practice 8B

1. 10, 20, 30, 40, 50, 60, 70, 80, 90, 100
2. 10, 20, 30, 40, 50, 60, 70, 80, 90, 100
3. 10, 20, 30, 40, 50, <u>60¢</u>
4. 10, 20, <u>30¢</u>
5. 10, 20, 30, 40, 50, 60, 70, <u>80¢</u>
6. 10, 20, 30, 40, 50, 60, 70,
80, 90, 100 pennies

Lesson Practice 8C

1. 10, 20, 30, 40, 50, 60, 70, 80, 90, 100
2. 10, 20, 30, 40, 50, 60, 70, 80, 90, 100
3. 10, 20, 30, 40, <u>50¢</u>
4. 10, 20, 30, 40, 50, 60, 70, 80, <u>90¢</u>
5. 10, 20, 30 pennies
6. 10, 20, 30, 40, 50, 60 dollars

Systematic Review 8D

1. 10, 20, 30, 40, 50, 60, 70, 80, 90, 100
2. 2, 4, 6, 8, 10, 12, 14, 16, 18, 20

3.
$$\begin{array}{r} 11 \\ +32 \\ \hline 43 \end{array}$$

4.
$$\begin{array}{r} 1 \\ 64 \\ +26 \\ \hline 90 \end{array}$$

5.
$$\begin{array}{r} 1 \\ 55 \\ +17 \\ \hline 72 \end{array}$$

6.
$$\begin{array}{r} 341 \\ +111 \\ \hline 452 \end{array}$$

7.
$$\begin{array}{r} 629 \\ +250 \\ \hline 879 \end{array}$$

8.
$$\begin{array}{r} 165 \\ +522 \\ \hline 687 \end{array}$$

7. $\begin{array}{r} 629 \\ +250 \\ \hline 879 \end{array}$

8. $\begin{array}{r} 165 \\ +522 \\ \hline 687 \end{array}$

9. $6 + \underline{1} = 7$

10. $6 + \underline{2} = 8$

11. $3 + \underline{0} = 3$

12. 10, $\underline{20}$ pennies

13. $(20) + (40) = (60)$ dollars
    $19 + 36 = 55$ dollars

14. $41 + 48 = 89$ birds

15. 9 dimes = 90¢
    95 pennies = 95¢
    95¢ > 90¢; so 95 pennies

Systematic Review 8E

1. 2, 4, 6, 8, 10, 12, 14, 16, 18, 20

2. 10, 20, 30, 40, 50, 60, 70, 80, 90, 100

3. $\begin{array}{r} 15 \\ +64 \\ \hline 79 \end{array}$

4. $\begin{array}{r} {\scriptstyle 1} \\ 13 \\ +28 \\ \hline 41 \end{array}$

5. $\begin{array}{r} {\scriptstyle 1} \\ 44 \\ +46 \\ \hline 90 \end{array}$

6. $\begin{array}{r} 175 \\ +114 \\ \hline 289 \end{array}$

7. $\begin{array}{r} 732 \\ +156 \\ \hline 888 \end{array}$

8. $\begin{array}{r} 244 \\ +234 \\ \hline 478 \end{array}$

9. $4 + \underline{2} = 6$

10. $6 + \underline{4} = 10$

11. $9 + \underline{9} = 18$

12. $6 + 2 = 8$
    $8 + 3 = 11$ toys

13. $2 + 5 = 7$ dimes
    10, 20, 30, 40, 50, 60, $\underline{70}$¢

14. $8 + \underline{4} = 12$

15. $242 + 157 = 399$ trees

Systematic Review 8F

1. 10, 20, 30, 40, 50, 60, 70, 80, 90, 100

2. 2, 4, 6, 8, 10, 12, 14, 16, 18, 20

3. $\begin{array}{r} {\scriptstyle 1} \\ 25 \\ +25 \\ \hline 50 \end{array}$

4. $\begin{array}{r} {\scriptstyle 1} \\ 19 \\ +\ 4 \\ \hline 23 \end{array}$

5. $\begin{array}{r} {\scriptstyle 1} \\ 38 \\ +15 \\ \hline 53 \end{array}$

6. $\begin{array}{r} 430 \\ +223 \\ \hline 653 \end{array}$

7. $\begin{array}{r} 805 \\ +\ 192 \\ \hline 997 \end{array}$

8. $\begin{array}{r} 317 \\ +651 \\ \hline 968 \end{array}$

9. $1 + \underline{7} = 8$

10. $9 + \underline{8} = 17$

11. $4 + \underline{1} = 5$

12. $56 + 39 = 95$ pieces

13. 2, 4, $\underline{6}$ bales

14. 8 dimes = 80¢
    80¢ > 8¢
    Chance has more

15. $2 + 2 = 4$
    $4 + 5 = 9$ hours

Lesson Practice 9A
1. 5, 10, 15, 20, 25, 30, 35, 40, 45, 50
2. 5, 10, 15, 20, 25, 30, 35, 40, 45, 50
3. 5, 10, 15, 20, 25, 30, 35, 40, 45, 50
4. 5, 10, 15, 20¢
5. 5, 10, 15, 20, 25 rooms
6. 5, 10, 15, 20, 25, 30, 35¢
7. 5, 10, 15, 20, 25, 30, 35, 40, 45 jellybeans

Lesson Practice 9B
1. 5, 10, 15, 20, 25, 30, 35, 40, 45, 50
2. 5, 10, 15, 20, 25, 30, 35, 40, 45, 50
3. 5, 10, 15, 20, 25, 30, 35, 40, 45, 50
4. 5, 10, 15, 20, 25¢
5. 5, 10, 15, 20, 25, 30 sides
6. 5, 10, 15, 20, 25, 30, 35, 40, 45¢
7. 5, 10, 15, 20, 25, 30, 35, 40 songs

Lesson Practice 9C
1. 5, 10, 15, 20, 25, 30, 35, 40, 45, 50
2. 5, 10, 15, 20, 25, 30, 35, 40, 45, 50
3. 5, 10, 15, 20, 25, 30, 35, 40, 45, 50
4. 5, 10, 15, 20, 25, 30, 35¢
5. 5, 10, 15 petals
6. 3 nickels = 15¢
   2 dimes = 20¢
   Bob has more
7. 5, 10, 15, 20¢

Systematic Review 9D
1. 5, 10, 15, 20, 25, 30, 35, 40, 45, 50
2. 10, 20, 30, 40, 50, 60, 70, 80, 90, 100
3. 63 + 7 = 70

4. 24 + 48 = 72
5. 15 + 44 = 59
6. 412 + 216 = 628
7. 203 + 302 = 505
8. 713 + 272 = 985
9. 6 + 2 = 8
10. 4 + 5 = 9
11. 7 + 7 = 14
12. 8 nickels = 40¢, so yes
13. 2, 4, 6, 8, 10 feet
14. 15 + 16 = 31 comic books
15. 6 + 2 = 8; 8 + 6 = 14 passengers

Systematic Review 9E
1. 5, 10, 15, 20, 25, 30, 35, 40, 45, 50
2. 2, 4, 6, 8, 10, 12, 14, 16, 18, 20
3. 27 + 33 = 60
4. 81 + 3 = 84
5. 36 + 14 = 50
6. 293 + 104 = 397
7. 645 + 321 = 966

8. 784
    +2 15
    999
9. 9+1=10
10. 3+6=9
11. 7+5=12
12. 5, 10, 15, 20, 25, 30,
    35, 40, 45, so 9 nickels
13. 25+16 = 41 times
14. 4+2=6
    6+8 = 14 runs
15. 5 nickels = 25¢
    4 dimes = 40¢
    40¢ > 25¢: so 4 dimes

Systematic Review 9F
1. 5, 10, 15, 20, 25, 30, 35, 40, 45, 50
2. 10, 20, 30, 40, 50, 60, 70, 80, 90, 100
3. 57
    +22
    79
4.   1
    74
    + 6
    80
5.   1
    24
    +18
    42
6. 680
    + 119
    799
7. 532
    +222
    754
8. 192
    +207
    399
9. 8+2=10
10. 7+8=15
11. 9+3=12

12. dime
13. penny
14. nickel

15. 314+322 = 636 dollars

Lesson Practice 10A
1. $1.66
    "one dollar and sixty-six cents"
2. $1.25
    "one dollar and twenty-five cents"
3. $2.19
    "two dollars and nineteen cents"
4. $1.30
    "one dollar and thirty cents"
5. 2 dollars, 6 dimes, and 2 pennies
    "two dollars and sixty-two cents"
6. 2 dollars and 5 pennies
    "two dollars and five cents"
7. 1 dollar, 9 dimes, and 6 pennies
    "one dollar and ninety-six cents"
8. 3 dollars, 1 dime, and 8 pennies
    "three dollars and eighteen cents"

BETA

Lesson Practice 10B

1. $3.52
   "three dollars and fifty-two cents"
2. $1.74
   "one dollar and seventy-four cents"
3. $2.46
   "two dollars and forty-six cents"
4. 4 dollars, 5 dimes, and 1 penny
   "four dollars and fifty-one cents"
5. 3 dollars and 6 dimes
   "three dollars and sixty cents"
6. 1 dollar, 8 dimes, and 1 penny
   "one dollar and eighty-one cents"
7. 2 dollars and 7 pennies
   "two dollars and seven cents"
8. two dollars and fifteen cents
   $2.15

Lesson Practice 10C

1. $4.01
   "four dollars and one cent"
2. $2.30
   "two dollars and thirty cents"
3. $1.49
   "one dollar and forty-nine cents"
4. 3 dollars, 2 dimes, and 3 pennies
   "three dollars and twenty-three cents"
5. 1 dollar and 8 pennies
   "one dollar and eight cents"
6. 4 dollars, 5 dimes, and 2 pennies
   "four dollars and fifty-two cents"
7. 2 dollars and 9 dimes
   "two dollars and ninety cents"
8. six dollars and seventy-three cents
   $6.73

Systematic Review 10D

1. $2.67
   "two dollars and sixty-seven cents"

2. 1 dollar, 4 dimes, and 8 pennies
   "one dollar and forty-eight cents"
3. 2 dollars, 7 dimes, and 3 pennies
   "two dollars and seventy-three cents"
4. 4 dollars and 5 pennies
   "four dollars and five cents"
5. 3 dollars and 6 dimes
   "three dollars and sixty cents"
6. 5, 10, 15, 20, $\underline{25}$¢
7. $\begin{array}{r} 1 \\ 49 \\ +\ 9 \\ \hline 58 \end{array}$
8. $\begin{array}{r} 311 \\ +238 \\ \hline 549 \end{array}$
9. $\begin{array}{r} 1 \\ 65 \\ +25 \\ \hline 90 \end{array}$
10. $7 + \underline{6} = 13$
11. $3 + \underline{2} = 5$
12. $4 + \underline{8} = 12$
13. five dollars and four cents = $5.04
14. $19 + 25 = 44$ rides
15. $4 + 4 = 8$
    $8 + 9 = 17$ cards

Systematic Review 10E

1. $3.20
   "three dollars and twenty cents"
2. 2 dollars, 3 dimes, and 1 penny
   "two dollars and thirty-one cents"
3. 4 dollars, 5 dimes, and 5 pennies
   "four dollars and fifty-five cents"
4. 1 dollar and 6 pennies
   "one dollar and six cents"
5. 3 dollars, 7 dimes, and 8 pennies
   "three dollars and seventy-eight cents"
6. 10, 20, 30, 40, 50, 60, 70, $\underline{80}$¢

7.
$$\begin{array}{r} 1 \\ 17 \\ +18 \\ \hline 35 \end{array}$$

8.
$$\begin{array}{r} 555 \\ +132 \\ \hline 687 \end{array}$$

9.
$$\begin{array}{r} 1 \\ 49 \\ +34 \\ \hline 83 \end{array}$$

10. 80

11. 10

12. 30

13. 5, 10, 15, 20, 25, 30,
35, 40, 45 = $.45

14. 5 + 2 = 7
7 + 2 = 9 roses

15. 30¢

Systematic Review10F

1. $1.08
"one dollar and eight cents"

2. 1 dollar, 1 dime, and 6 pennies
"one dollar and sixteen cents"

3. 3 dollars and 9 pennies
"three dollars and nine cents"

4. 2 dollars, 6 dimes, and 5 pennies
"two dollars and sixty-five cents"

5. 4 dollars and 7 dimes
"four dollars and seventy cents"

6. 5, 10, 15¢

7.
$$\begin{array}{r} 92 \\ +\ 4 \\ \hline 96 \end{array}$$

8.
$$\begin{array}{r} 337 \\ +202 \\ \hline 539 \end{array}$$

9.
$$\begin{array}{r} 1 \\ 61 \\ +29 \\ \hline 90 \end{array}$$

10. 10 > 7

11. 8 = 8

12. 27 < 72

13. $8.69

14. 3 + 4 = 7
7 + 8 = 15 miles

15. 29 + 18 = 47 miles

Lesson Practice 11A

1. 200

2. 200

3. 400

4. done

5.
$$\begin{array}{r} (600) \\ +(200) \\ \hline (800 \end{array} \quad \begin{array}{r} 11 \\ 628 \\ +175 \\ \hline 803 \end{array}$$

6.
$$\begin{array}{r} (400) \\ +(300) \\ \hline (700) \end{array} \quad \begin{array}{r} 11 \\ 359 \\ +254 \\ \hline 613 \end{array}$$

7.
$$\begin{array}{r} (500) \\ +(200) \\ \hline (700) \end{array} \quad \begin{array}{r} 1 \\ 537 \\ +233 \\ \hline 770 \end{array}$$

8.
$$\begin{array}{r} (200) \\ +(500) \\ \hline (700) \end{array} \quad \begin{array}{r} 11 \\ 168 \\ +452 \\ \hline 620 \end{array}$$

9.
$$\begin{array}{r} (100) \\ +(100) \\ \hline (200) \end{array} \quad \begin{array}{r} 11 \\ 123 \\ +\ 88 \\ \hline 211 \end{array}$$

10.
$$\begin{array}{r} (700) \\ +(100) \\ \hline (800) \end{array} \quad \begin{array}{r} 11 \\ 676 \\ +145 \\ \hline 821 \end{array}$$

11.
$$\begin{array}{r} (300) \\ +(300) \\ \hline (600) \end{array} \quad \begin{array}{r} 11 \\ 299 \\ +311 \\ \hline 610 \end{array}$$

12. 124 + 176 = 300 lights

Lesson Practice 11B

1. 500
2. 500
3. 600

4.
```
     1
(400)   359
+(100)  +126
(500)   485
```

5.
```
     1
(100)   138
+(200)  +212
(300)   350
```

6.
```
(200)   157
+(100)  +142
(300)   299
```

7.
```
(200)   227
+(000)  + 39
(200)   266
```

8.
```
     1
(400)   449
+(100)  +137
(500)   586
```

9.
```
     1
(200)   235
+(100)  +145
(300)   380
```

10.
```
     1
(100)   109
+(200)  +207
(300)   316
```

11.
```
     1
(400)   416
+(300)  +329
(700)   745
```

12. 123+169 = 292 pages

Lesson Practice 11C

1. 500
2. 100
3. 300

4.
```
       1
(200)   217
+(300)  +324
(500)   541
```

5.
```
     1
(300)   266
+(000)  + 18
(300)   284
```

6.
```
(100)   134
+(400)  +365
(500)   499
```

7.
```
     1
(100)   119
+(200)  +207
(300)   326
```

8.
```
       11
(600)   555
+(300)  +348
(900)   903
```

9.
```
     1
(800)   806
+(100)  +106
(900)   912
```

10.
```
     1
(100)   119
+(200)  +217
(300)   336
```

11.
```
       11
(200)   248
+(300)  +252
(500)   500
```

12. 263+179 = 442 miles

Systematic Review 11D

1. 800
2. 100

3.
```
    1
   806
 + 106
 9 12
```

4.
```
   11
   248
 +252
   500
```

5.
```
  1
  337
+172
  509
```

6.
```
  1
  54
+28
  82
```

7.
```
  1
  53
+37
  90
```

8.
```
  1
  18
+29
  47
```

9. $1-1=0$
10. $10-2=8$
11. $8-1=7$
12. $3-0=3$
13. $4-3=1$
14. $6-2=4$
15. $5-4=1$
16. $8-2=6$
17. 2, 4, 6, 8, 10, 12, 14, 16, 18, 20
18. $5.26
19. $55+78=133$ miles
20. $145+$56=$201

6.
```
  1
  28
+38
  66
```

7.
```
  1
  65
+35
  100
```

8.
```
  1
  58
+42
  100
```

9. $4-2=2$
10. $7-2=5$
11. $3-1=2$
12. $11-2=9$
13. $6-5=1$
14. $8-0=8$
15. $10-9=1$
16. $9-2=7$
17. 5, 10, 15, 20, 25, 30, 35, 40, 45, 50
18. $17+9=26$ inches
19. $138+256=394$ penguins
20. $5-2=3$ eggs

Systematic Review 11E

1. 400
2. 200

3.
```
  11
  235
+365
  600
```

4.
```
  300
+409
  709
```

5.
```
  1
  249
+132
  381
```

Systematic Review 11F

1. 500
2. 700

3.
```
  1
  429
+266
  695
```

4.
```
   1
  10 1
+  89
  190
```

5.
```
  1
  238
+243
  481
```

6.
```
  1
  92
+  8
  100
```

7.
```
   1
   48
 +32
   80
```

8.
```
   1
   63
 +27
   90
```

9. $5 - 3 = 2$

10. $10 - 2 = 8$

11. $7 - 5 = 2$

12. $6 - 2 = 4$

13. $9 - 7 = 2$

14. $8 - 2 = 6$

15. $10 - 8 = 2$

16. $11 - 9 = 2$

17. 10, 20, 30, 40, 50, 60, 70, 80, 90, 100

18. $(20) + (50) = (70)$ pieces

19. $(500) + (300) = (800)$
    $512 + 345 = 857$ miles

20. $\$.08 - \$.06 = \$.02$

7.
```
  $1.00
 + .75
  $1.75
```

8.
```
  $2.03
 +1.90
  $3.93
```

9.
```
     1
  $8.75
 + .80
  $9.55
```

10. $\$5.25 + \$3.38 = \$8.63$

11. $\$2.63 + \$5.50 = \$8.13$

12. $\$2.99 + \$3.61 = \$6.60$

Lesson Practice 12A

1. done

2.
```
    11
  $7.09
 +1.92
  $9.01
```

3.
```
  $3.33
 +1.44
  $4.77
```

4.
```
     1
  $6.50
 +2.77
  $9.27
```

5.
```
  $4.00
 +2.51
  $6.51
```

6.
```
     1
  $5.19
 +1.38
  $6.57
```

Lesson Practice 12B

1.
```
     1
  $7.65
 + .60
  $8.25
```

2.
```
     1
  $6.31
 +1.29
  $7.60
```

3.
```
     1
  $5.83
 + .24
  $6.07
```

4.
```
     1
  $3.19
 + .90
  $4.09
```

5.
```
  $2.00
 + .98
  $2.98
```

6.
```
  $1.03
 +1.25
  $2.28
```

7.
```
     1
  $3.72
 +4.08
  $7.80
```

8.
```
    11
  $1.99
 +1.82
  $3.81
```

9.
```
  1 1
 $2.87
+6.89
 $9.76
```

10. $3.45 + $1.99 = $5.44

11. $5.55 + $2.15 = $7.70

12. $6.34 + $2.95 = $9.29

Lesson Practice 12C

1.
```
  1
 $2.13
+1.92
 $4.05
```

2.
```
  1
 $4.71
+1.36
 $6.07
```

3.
```
  1
 $6.41
+ .39
 $6.80
```

4.
```
 $5.00
+2.50
 $7.50
```

5.
```
  1
 $6.63
+2.44
 $9.07
```

6.
```
  1
 $7.35
+1.05
 $8.40
```

7.
```
  1
 $1.63
+ .72
 $2.35
```

8.
```
  1 1
 $4.99
+3.79
 $8.78
```

9.
```
  1
 $6.33
+2.91
 $9.24
```

10. $5.10 + $3.91 = $9.01

11. $2.50 + $4.50 = $7.00

12. $3.72 + $3.68 = $7.40

$7.40 > $6.00; yes

Systematic Review 12D

1.
```
  1
 $1.66
+4.08
 $5.74
```

2.
```
  1
 $3.09
+2.56
 $5.65
```

3.
```
  1
 $3.57
+2.62
 $6.19
```

4.
```
  1 1
  422
 +389
  811
```

5.
```
  1
  19
 +16
  35
```

6.
```
  1
  17
 +25
  42
```

7. 12 − 9 = 3

8. 18 − 9 = 9

9. 9 − 9 = 0

10. 14 − 9 = 5

11. 17 − 9 = 8

12. 13 − 9 = 4

13. 16 − 9 = 7

14. 15 − 9 = 6

15. $2.35

"two dollars and thirty-five cents"

16. 11 − 9 = 2 gifts

17. (50) + (60) = (110)

46 + 63 = 109 chairs

18. $2.78 + $1.19 = $3.97

Systematic Review 12E

1.
```
  1 1
$1.68
+4.77
$6.45
```

2.
```
  1 1
$4.56
+4.44
$9.00
```

3.
```
  1
$2.63
+ .51
$3.14
```

4.
```
  1
 684
+122
 806
```

5.
```
 1
 62
+29
 91
```

6.
```
 1
 83
+ 7
 90
```

7. $11 - 8 = 3$
8. $9 - 8 = 1$
9. $17 - 8 = 9$
10. $12 - 8 = 4$
11. $14 - 8 = 6$
12. $13 - 8 = 5$
13. $15 - 8 = 7$
14. $16 - 8 = 8$
15. $1.26
   "one dollar and twenty-six cents"
16. $13 - 8 = 5$
   5 dimes = 50¢
17. $(300) + (200) = (500)$
   $267 + 197 = 464$ chocolates
18. 3 dimes = 30¢
   5 nickels = 25¢
   30¢ + 25¢ = 55¢
   $.55

Systematic Review 12F

1.
```
  1 1
$2.78
+6.58
$9.36
```

2.
```
  1
$3.52
+1.77
$5.29
```

3.
```
$8.91
+ .05
$8.96
```

4.
```
  1 1
 379
+264
 643
```

5.
```
 1
 54
+18
 72
```

6.
```
 1
 47
+ 9
 56
```

7. $12 - 6 = 6$
8. $13 - 8 = 5$
9. $10 - 5 = 5$
10. $14 - 7 = 7$
11. $12 - 8 = 4$
12. $17 - 8 = 9$
13. $8 - 4 = 4$
14. $6 - 3 = 3$
15. $.54
   "fifty-four cents"
16. $16 - 8 = 8$ birds
17. 3 nickels = 15¢
   15¢ + 5¢ = 20¢
   $.20
18. $35 + 17 = 52$ minutes

Lesson Practice 13A

1. $2+8+5=15$
2. $6+3+4=13$
3. $5+4+5=14$
4. $8+2+6+1=17$
5. $6+2+4+7=19$
6. $3+7+5+5=20$
7. $40+50+10=100$
8. $30+70+40=140$
9. $51+59+20=130$
10. $9+1+8+2=20$
11. $6+1+8+4+2=21$
12. $4+5+6+5=20$ presents

Lesson Practice 13B

1. $6+4+9=19$
2. $9+5+5=19$
3. $7+2+3=12$
4. $8+5+4+2=19$
5. $9+1+7+3=20$
6. $1+2+3+9=15$
7. $60+40+20=120$
8. $20+60+80=160$
9. $43+27+61=131$
10. $8+2+3+7+2=22$
11. $4+9+6+2+1=22$
12. $8+2+3+3=16$ laps

Lesson Practice 13C

1. $2+8+7=17$
2. $5+1+5=11$
3. $4+3+3=10$
4. $2+8+2+3=15$
5. $3+8+7+2=20$
6. $9+5+6+1=21$
7. $80+50+10+10=150$
8. $40+60+4+2=106$
9. $84+26+17+23=150$
10. $6+7+3+4+6=26$
11. $5+5+4+3+7=24$
12. $11+14+5+10+4=44$ books

Systematic Review 13D

1. $3+7+6=16$
2. $3+9+1+7=20$
3. $42+64+82+2=190$
4. $\begin{array}{r} 1\ 1 \\ \$2.85 \\ +6.56 \\ \hline \$9.41 \end{array}$
5. $\begin{array}{r} 11 \\ 149 \\ +273 \\ \hline 422 \end{array}$
6. $\begin{array}{r} 1 \\ 14 \\ +\ \ 9 \\ \hline 23 \end{array}$
7. $10-8=2$
8. $10-4=6$
9. $9-6=3$
10. $10-7=3$
11. $9-2=7$
12. $9-4=5$
13. $9-1=8$
14. $10-6=4$
15. 10, 20, 30, 40, 50, 60, 70, 80, 90, 100
16. triangle
17. $13+15+11=39$ patients
18. $\$6.18+\$2.00=\$8.18$
$\$8.18 < \$9.00$; no

Systematic Review 13E

1. $2+4+2=8$
2. $6+3+8+2=19$
3. $10+10+20+20=60$
4. $\begin{array}{r} 1\ 1 \\ \$3.46 \\ +2.54 \\ \hline \$6.00 \end{array}$
5. $\begin{array}{r} 100 \\ +278 \\ \hline 378 \end{array}$

6.
$$\begin{array}{r} 1 \\ 79 \\ +88 \\ \hline 167 \end{array}$$

7. $10 - 2 = 8$

8. $10 - 5 = 5$

9. $9 - 5 = 4$

10. $9 - 7 = 2$

11. $10 - 3 = 7$

12. $9 - 8 = 1$

13. $9 - 3 = 6$

14. $10 - 4 = 6$

15. 2, 4, 6, 8, 10, 12, 14, 16, 18, 20

16. square, rectangle

17. $35 + 22 + 12 = 69$ cones

18. 9 nickels = 45¢

    $45¢ + 7¢ = 52¢$

    $.52

Systematic Review 13F

1. $5 + 5 + 2 = 12$

2. $5 + 6 + 4 + 7 = 22$

3. $23 + 37 + 23 + 12 = 95$

4.
$$\begin{array}{r} \$4.36 \\ +1.22 \\ \hline \$5.58 \end{array}$$

5.
$$\begin{array}{r} 1 \\ 116 \\ + 68 \\ \hline 184 \end{array}$$

6.
$$\begin{array}{r} 1 \\ 16 \\ +44 \\ \hline 60 \end{array}$$

7. $17 - 8 = 9$

8. $14 - 7 = 7$

9. $10 - 3 = 7$

10. $9 - 5 = 4$

11. $15 - 9 = 6$

12. $9 - 7 = 2$

13. $12 - 8 = 4$

14. $16 - 9 = 7$

15. 5, 10, 15, 20, 25, 30, 35, 40, 45, 50

16. rectangle

17. triangle

18. square

19. 6 nickels = 30¢

    5 dimes = 50¢

    $30¢ + 50¢ + 4¢ = 84¢$

    $.84

20. $33 + 49 = 82$ raisins

Lesson Practice 14A

1. sides: 2"

   base: 3"

2. 4"

3. 5"

4. 3"

5. sides: 2"

   top and bottom: 3"

6. 1 foot

Lesson Practice 14B

1. all sides: 1"

2. 2"

3. 7"

4. 6"

5. all sides: 2"

6. 12

Lesson Practice 14C

1. sides: 1"
   top and bottom: 3"
2. 5"
3. 1"
4. 4"
5. all sides: 3"
6. $12" + 12" = 24"$

Systematic Review 14D

1. 3"
2. 7"
3.
$$
\begin{array}{r}
11 \\
\$1.77 \\
+2.78 \\
\hline
\$4.55
\end{array}
$$
4.
$$
\begin{array}{r}
1 \\
118 \\
+122 \\
\hline
240
\end{array}
$$
5.
$$
\begin{array}{r}
23 \\
+24 \\
\hline
47
\end{array}
$$
6. $1+2+3+7+8+9 = 30$
7. $6+6+4+3+7 = 26$
8. $6 = 6$
9. $6 < 8$
10. $7-3 = 4$
11. $8-5 = 3$
12. $7-4 = 3$
13. $8-3 = 5$
14. $9-5 = 4$
15. $15-9 = 6$
16. three
17. $6-3 = 3$ dimes
    3 dimes $= 30$¢ or $.30
18. $6+5+10+4 = 25$ pieces

Systematic Review 14E

1. 6"
2. 2"
3.
$$
\begin{array}{r}
11 \\
\$3.63 \\
+2.77 \\
\hline
\$6.40
\end{array}
$$
4.
$$
\begin{array}{r}
1 \\
451 \\
+181 \\
\hline
632
\end{array}
$$

handwritten:
$$
\begin{array}{r}
1 \\
452 \\
+181 \\
\hline
633
\end{array}
$$
← In books

5.
$$
\begin{array}{r}
1 \\
89 \\
+87 \\
\hline
176
\end{array}
$$
6. $3+3+7+3+2 = 18$
7. $10+2+10+2 = 24$
8. $4 > 3$
9. $11-7 = 4$
10. $13-7 = 6$
11. $8-3 = 5$
12. $16-7 = 9$
13. $12-7 = 5$
14. $15-7 = 8$
15. four
16. $256+289 = 545$ miles
17. $3+4+1+2 = 10$ calls
18. $12+12+12 = 36"$

Systematic Review 14F

1. 4"
2. 5"
3.
$$
\begin{array}{r}
1 \\
\$1.56 \\
+2.38 \\
\hline
\$3.94
\end{array}
$$
4.
$$
\begin{array}{r}
1 \\
419 \\
+419 \\
\hline
838
\end{array}
$$
5.
$$
\begin{array}{r}
1 \\
39 \\
+42 \\
\hline
81
\end{array}
$$

6. $5+5+5+5 = 20$

7. $3+5+10 = 18$

8. $5 < 8$

9. $11-6 = 5$

10. $14-6 = 8$

11. $15-6 = 9$

12. $13-6 = 7$

13. $15-7 = 8$

14. $7-4 = 3$

15. four

16. 7 dimes $= 70$¢

    3 nickels $= 15$¢

    70¢$+15$¢$ = 85$¢ or $.85

17. $3+3 = 6$ pages read

    $13-6 = 7$ pages left

18. $56+35 = 91$ miles

Lesson Practice 15A

1. done

2. square

   $6+6+6+6 = 24$"

3. triangle

   $3+4+5 = 12$"

4. rectangle

   $7+2+7+2 = 18$"

Lesson Practice 15B

1. rectangle

   $3+8+3+8 = 22$"

2. square

   $2+2+2+2 = 8$"

3. triangle

   $6+8+10 = 24$"

4. rectangle

   $4+6+4+6 = 20$"

Lesson Practice 15C

1. square

   $9+9+9+9 = 36$"

2. rectangle

   $2+5+2+5 = 14$"

3. square

   $7+7+7+7 = 28$"

4. triangle

   $4+7+10 = 21$"

Systematic Review 15D

1. triangle

   $6+6+7 = 19$"

2. rectangle

   $10+15+10+15 = 50$"

3. $$\begin{array}{r} 1 \\ \$2.49 \\ +1.32 \\ \hline \$3.81 \end{array}$$

4. $$\begin{array}{r} 300 \\ +409 \\ \hline 709 \end{array}$$

5. $$\begin{array}{r} 1 \\ 27 \\ +25 \\ \hline 52 \end{array}$$

6. $11-5 = 6$

7. $12-4 = 8$

8. $13-5 = 8$

9. $11-3 = 8$

10. $12-5 = 7$

11. $11-4 = 7$

12. $10-7 = 3$ years

13. $53+8 = 61$ papers

14. $12+12 = 24$"

15. 9 nickels $= 45$¢

    1 dime $= 10$¢

    45¢$+10$¢$ = 55$¢ or $.55

Systematic Review 15E

1. square
   $6+6+6+6 = 24"$
2. rectangle
   $5+12+5+12 = 34"$
3. $\begin{array}{r} {}^{1} \\ \$4.28 \\ +1.65 \\ \hline \$5.93 \end{array}$
4. $\begin{array}{r} {}^{11} \\ 285 \\ +156 \\ \hline 441 \end{array}$
5. $\begin{array}{r} {}^{1} \\ 45 \\ +55 \\ \hline 100 \end{array}$
6.
7	9	16
3	5	8
10	14	24
7.
6	4	10
9	8	17
15	12	27
8. $14-5 = 9$
9. $12-3 = 9$
10. $6-1 = 5$
11. $13-4 = 9$
12. $4+4+4+4 = 16"$
13. $31+31+28 = 90$ days
14. $12+12+12+12 = 48"$

Systematic Review 15F

1. triangle
   $7+8+9 = 24"$
2. rectangle
   $11+18+11+18 = 58"$
3. $\begin{array}{r} {}^{1} \\ \$2.48 \\ +1.29 \\ \hline \$3.77 \end{array}$
4. $\begin{array}{r} {}^{11} \\ 183 \\ +\;\;77 \\ \hline 260 \end{array}$

5. $\begin{array}{r} 63 \\ +26 \\ \hline 89 \end{array}$
6. 2, 4, 6, 8, 10, 12, 14, 16, 18, 20
7. 5, 10, 15, 20, 25, 30, 35, 40, 45, 50
8. $9-6 = 3$
9. $11-7 = 4$
10. $12-5 = 7$
11. $13-9 = 4$
12. $11-5 = 6$ guests
13. $6+8+10 = 24'$ (24 feet)
14. $9+9 = 18$ runs

Lesson Practice 16A

1. done
2. 35,361
   "thirty-five thousand,
   three hundred sixty-one"
3. 785,892
   "seven hundred eighty-five thousand,
   eight hundred ninety-two"
4. 265,143
   "two hundred sixty-five thousand,
   one hundred forty-three"
5. 6,237
   "six thousand, two hundred thirty-seven"
6. done
7. $2,000+300+50+6$
8. done
9. 1,542
10. done
11. $\begin{array}{r} {}^{1} \\ 915 \\ +436 \\ \hline 1,351 \end{array}$
12. $\begin{array}{r} {}^{1} \\ 381 \\ +727 \\ \hline 1,108 \end{array}$

Lesson Practice 16B

1. 2,794
"two thousand, seven hundred ninety-four"
2. 16,322
"sixteen thousand,
three hundred twenty-two"
3. 651,741
"six hundred fifty-one thousand,
seven hundred forty-one"
4. 536,583
"five hundred thirty-six thousand,
five hundred eighty-three"
5. 2,549
"two thousand, five hundred forty-nine"
6. $40,000 + 1,000 + 400 + 50 + 6$
7. $200,000 + 30,000 + 8,000 + 100 + 90 + 9$
8. 3,121
9. 45,616
10.
```
   1
  593
 +551
 1,144
```
11.
```
   1
  876
 +431
 1,307
```
12.
```
  967
 +202
 1,169
```

Lesson Practice 16C

1. 1,224
"one thousand,
two hundred twenty-four"
2. 43,638
"forty-three thousand,
six hundred thirty-eight"
3. 247,500
"two hundred forty-seven thousand,
five hundred"

4. 122,472
"one hundred twenty-two thousand,
four hundred seventy-two"
5. 7,294
"seven thousand,
two hundred ninety-four"
6. $50,000 + 6,000 + 600 + 40 + 4$
7. $3,000 + 200 + 50 + 6$
8. 1,838
9. 33,230
10.
```
   453
 +7 14
 1, 167
```
11.
```
   11
   345
  +978
  1,323
```
12.
```
  7 16
 +563
 1,279
```

Systematic Review 16D

1. 4,819
"four thousand,
eight hundred nineteen"
2. 57,284
"fifty-seven thousand,
two hundred eighty-four"
3. rectangle
$11+32+11+32 = 86$"
4.
```
   90 1
  +850
  1,75 1
```
5.
```
    1
   534
  +673
  1,207
```
6.
```
  $4. 12
  +4.7 1
  $8.83
```
7. $17+23+55 = 95$
8. $10 - 6 = 4$
9. $15 - 8 = 7$

10. $14 - 6 = 8$
11. $6 - 3 = 3$
12. $12 - 7 = 5$
13. $6 - 4 = 2$
14. $\$48 + \$32 + \$21 = \$101$
15. $12 + 12 + 12 = 36"$
16. $16 - 8 = 8$ years

Systematic Review 16E

1. 6,211
   "six thousand, two hundred eleve
2. 28,616
   "twenty-eight thousand,
   six hundred sixteen"
3. triangle
   $11 + 12 + 13 = 36"$
4. $\begin{array}{r} 1 \\ 559 \\ +524 \\ \hline 1,083 \end{array}$
5. $\begin{array}{r} 1 \\ 943 \\ +475 \\ \hline 1,418 \end{array}$
6. $\begin{array}{r} 1 \\ \$\ 3.37 \\ +\ 8.29 \\ \hline \$11.66 \end{array}$
7. $65 + 28 + 45 = 138$
8. $9 - 3 = 6$
9. $8 - 5 = 3$
10. $14 - 8 = 6$
11. $11 - 4 = 7$
12. $16 - 7 = 9$
13. $7 - 1 = 6$
14. $\$(200) + \$(200) = \$(400)$
    $\$198 + \$242 = \$440$
15. 8 dimes $= 80¢$
    $80¢ - 30¢ = 50¢$ or $.50$
16. $9 + 7 + 9 + 7 = 32'$

Systematic Review 16F

1. 7,854
   "seven thousand, eight hundred fifty-four"
2. 815,231
3. square
   $6 + 6 + 6 + 6 = 24"$
4. $\begin{array}{r} 624 \\ +413 \\ \hline 1,037 \end{array}$
5. $\begin{array}{r} 426 \\ +873 \\ \hline 1,299 \end{array}$
6. $\begin{array}{r} 1 \\ \$1.32 \\ +3.38 \\ \hline \$4.70 \end{array}$
7. $71 + 53 + 11 = 135$
8. $8 - 7 = 1$
9. $11 - 8 = 3$
10. $14 - 9 = 5$
11. $9 - 4 = 5$
12. $13 - 5 = 8$
13. $8 - 3 = 5$
14. $(500) + (500) = (1,000)$
    $512 + 471 = 983$ cars
15. $10 - 7 = 3$ apples
16. 5 nickels $= 25¢$
    4 dimes $= 40¢$
    $40¢ > 25¢$; so 4 dimes

Lesson Practice 17A

1. 5,000
2. 1,000
3. 90,000
4. 50,000
5. done
6. $\begin{array}{r} 111 \\ 4,859 \quad (5,000) \\ +2,444 \quad +(2,000) \\ \hline 7,303 \quad (7,000) \end{array}$

7.
```
   1
  9,253      (9,000)
 +7,845    + (8,000)
 17,098     (17,000)
```

8.
```
   1
  7,132      (7,000)
 +1,186    + (1,000)
  8,318      (8,000)
```

9.
```
   1  1
  3,624      (4,000)
 +4,418    + (4,000)
  8,042      (8,000)
```

10.
```
   1 1 1
  2,852      (3,000)
 +3,149    + (3,000)
  6,001      (6,000)
```

11. 3,152 + 7,321 = 10,473 miles

12. 5,232 + 3,765 = 8,997 fish

Lesson Practice 17B

1. 7,000
2. 4,000
3. 50,000
4. 20,000

5.
```
   1 1
  5,242      (5,000)
 +3,765    + (4,000)
  9,007      (9,000)
```

6.
```
    1
  9,287      (9,000)
 +1,321    + (1,000)
 10,608     (10,000)
```

7.
```
   1 1
  6,463      (6,000)
 +8,765    + (9,000)
 15,228     (15,000)
```

8.
```
    1
  7,214      (7,000)
 +1,108    + (1,000)
  8,322      (8,000)
```

9.
```
   1  1
  2,817      (3,000)
 +9,236    + (9,000)
 12,053     (12,000)
```

10.
```
   1 1
  3,680      (4,000)
 +3,384    + (3,000)
  7,064      (7,000)
```

11. 5,740 + 4,291 = 10,031 plants

12. 4,987 + 3,732 = 8,719 stars

Lesson Practice 17C

1. 1,000
2. 6,000
3. 80,000
4. 10,000

5.
```
  9,413      (9,000)
 +1,245    + (1,000)
 10,658     (10,000)
```

6.
```
    1
  9,287      (9,000)
 +7,491    + (7,000)
 16,778     (16,000)
```

7.
```
   1 1 1
  5,486      (5,000)
 +4,528    + (5,000)
 10,014     (10,000)
```

8.
```
  9,025      (9,000)
 +3,354    + (3,000)
 12,379     (12,000)
```

9.
```
  7,513      (8,000)
 +7,254    + (7,000)
 14,767     (15,000)
```

10.
```
   1 1
  1,890      (2,000)
 +3,672    + (4,000)
  5,562      (6,000)
```

11. $5,486 + $1,194 = $6,680

12. 8,972 + 9,221 = 18,193 people

Systematic Review 17D

1. 3,000

2. 1,000

3.
$$\begin{array}{r} {\scriptstyle 1\,1\,1} \\ 6,788 \\ +2,467 \\ \hline 9,255 \end{array}$$

4.
$$\begin{array}{r} {\scriptstyle 1\,1} \\ 2,355 \\ +1,672 \\ \hline 4,027 \end{array}$$

5.
$$\begin{array}{r} \$\ 8.42 \\ +\ 3.21 \\ \hline \$11.63 \end{array}$$

6. 3,188

   "three thousand, one hundred eighty-eight"

7. $8+16+8+16 = 48"$

8. $14-7 = 7$

9. $11-6 = 5$

10. $7-3 = 4$

11. $12-3 = 9$

12. $9-5 = 4$

13. $8-4 = 4$

14. $5+5 = 10$ eggs picked up

    $10-3 = 7$ eggs left

15. $296+316 = 612$ feet

16. $\$16+\$18+\$9 = \$43$

Systematic Review 17E

1. 40,000

2. 80,000

3.
$$\begin{array}{r} {\scriptstyle 1} \\ 1,476 \\ +7,813 \\ \hline 9,289 \end{array}$$

4.
$$\begin{array}{r} 1,621 \\ +4,157 \\ \hline 5,778 \end{array}$$

5.
$$\begin{array}{r} {\scriptstyle 1} \\ \$7.16 \\ +2.79 \\ \hline \$9.95 \end{array}$$

6. 5,400

   "five thousand four hundred"

7. $15+15+15+15 = 60'$

8. $12-4 = 8$

9. $10-3 = 7$

10. $7-4 = 3$

11. $3-2 = 1$

12. $15-6 = 9$

13. $9-3 = 6$

14. 3 dimes $= 30¢$

    $30¢ + 4¢ = 34¢$ or $\$.34$

15. $75+108 = 183$ items

16. $1,575+1,892 = 3,467$ mosquitoes

Systematic Review 17F

1. 1,000

2. 9,000

3.
$$\begin{array}{r} {\scriptstyle 1} \\ 7,438 \\ +9,114 \\ \hline 16,552 \end{array}$$

4.
$$\begin{array}{r} {\scriptstyle 1} \\ 6,408 \\ +4,379 \\ \hline 10,787 \end{array}$$

5.
$$\begin{array}{r} {\scriptstyle 1} \\ \$2.56 \\ +1.24 \\ \hline \$3.80 \end{array}$$

6. 131,528

   "one hundred thirty-one thousand, five hundred twenty-eight"

7. $9+12+15 = 36'$

8. $13-4 = 9$

9. $11-3 = 8$

10. $14-5 = 9$

11. $10-9 = 1$

12. $6-0 = 6$

13. $8-8 = 0$

14. $12-7 = 5$ people

15. $(200)+(300) = (500)$

    $235+256 = 491$ miles

16. $10+12+18 = 40$ flowers

Lesson Practice 18A

1. done

2.
```
  22
  294    (300)
  187    (200)
  306    (300)
 +813  + (800)
 1,600   (1,600)
```

3.
```
   22
  493    (500)
  215    (200)
  485    (500)
  324    (300)
 +106  + (100)
 1,623   (1,600)
```

4.
```
   22
  613    (600)
   97    (100)
  452    (500)
  879    (900)
 + 30  + (000)
 2,071   (2,100)
```

5.
```
   12
  113    (100)
  251    (300)
  345    (300)
  355    (400)
 +427  + (400)
 1,491   (1,500)
```

6.
```
   22
  546    (500)
  120    (100)
  309    (300)
  675    (700)
 +481  + (500)
 2,131   (2,100)
```

7. $(100) + $(500) + $(200) = $(800)
   $127 + $475 + $225 = $827

8. (400) + (200) + (200) = (800)
   390 + 240 + 152 = 782 miles

Lesson Practice 18B

1.
```
   21
  264    (300)
   85    (100)
  624    (600)
 +945  + (900)
 1,918   (1,900)
```

2.
```
   21
  172    (200)
  261    (300)
  527    (500)
 +446  + (400)
 1,406   (1,400)
```

3.
```
   22
  933    (900)
   58    (100)
  361    (400)
  159    (200)
 +542  + (500)
 2,053   (2,100)
```

4.
```
   21
  142    (100)
  206    (200)
  860    (900)
  462    (500)
 +553  + (600)
 2,223   (2,300)
```

5.
```
   22
  321    (300)
   39    (000)
  686    (700)
  452    (500)
 +152  + (200)
 1,650   (1,700)
```

6.
```
   22
  214    (200)
  596    (600)
  473    (500)
  527    (500)
 +802  + (800)
 2,612   (2,600)
```

7. 208 + 316 + 365 = 889 gallons

8. 137 + 122 + 101 + 150 = 510 students

Lesson Practice 18C

```
        12
1.     649      (600)
       536      (500)
        31      (000)
      +224    + (200)
     1,440     (1,300)
```

```
        12
2.     714      (700)
       746      (700)
       419      (400)
      +652    + (700)
     2,531     (2,500)
```

```
        22
3.     328      (300)
       530      (500)
       356      (400)
       432      (400)
      +456    + (500)
     2,102     (2,100)
```

```
        22
4.     219      (200)
       820      (800)
       479      (500)
       381      (400)
      +161    + (200)
     2,060     (2,100)
```

```
        12
5.     562      (600)
       520      (500)
        38      (000)
       659      (700)
      +613    + (600)
     2,392     (2,400)
```

```
        22
6.     247      (200)
       254      (300)
       416      (400)
       252      (300)
      +547    + (500)
     1,716     (1,700)
```

7. $553+334+129 = 1,016$ flakes
8. $302+148+447 = 897$ cans

Systematic Review18D

```
        22
1.     296
       308
       742
       866
      +314
     2,526
```

```
        12
2.     521
        52
       624
       546
      + 38
     1,781
```

```
        21
3.     273
       951
       550
       339
      +282
     2,395
```

4. 7,821; "seven thousand, eight hundred twenty-one"
5. 350,000
6. $2-0=2$
7. $4-1=3$
8. $6-2=4$
9. $5-5=0$
10. done
11. done
12. –
13. +
14. 2, 4, 6, 8, 10, 12, 14, 16, 18, 20
15. $51+29+19 = 99$ guests
16. $68+89+132 = 289$ flowers

Systematic Review18E

```
        21
1.     457
       561
       451
       631
      + 29
     2,129
```

2.
```
   22
  873
  265
  314
  247
+936
2,635
```

3.
```
   22
  804
  246
  531
  382
+639
2,602
```

4. 81,246
"eighty-one thousand, two hundred forty-six"

5. 652,693

6. $12 - 9 = 3$

7. $15 - 9 = 6$

8. $11 - 9 = 2$

9. $17 - 9 = 8$

10. +

11. −

12. −

13. +

14. 5, 10, 15, 20, 25, 30, 35, 40, 45, 50

15. $12 + 12 + 12 + 12 + 12 + 12 = 72$"

16. 3 nickels = 15¢; 4 dimes = 40¢
15¢ + 40¢ = 55¢ or $.55

Systematic Review18F

1.
```
   11
  741
  405
  401
  586
+364
2,497
```

2.
```
   22
  952
  237
  171
  603
+459
2,422
```

3.
```
   21
  631
  834
  841
  272
+234
2,812
```

4. 637,531
"six hundred thirty-seven thousand, five hundred thirty-one"

5. 45,727

6. $11 - 8 = 3$

7. $14 - 8 = 6$

8. $17 - 8 = 9$

9. $13 - 8 = 5$

10. −

11. −

12. +

13. −

14. 10, 20, 30, 40, 50, 60, 70, 80, 90, 100

15. $87 + 48 + 211 = 346$ pages

16. $16 - 7 = 9$ years

Lesson Practice 19A

1. done

2.
```
   11
  7,132    (7,000)
  5,333    (5,000)
+1,186  + (1,000)
 13,651   (13,000)
```

3.
```
   111
  2,852    (3,000)
  4,263    (4,000)
+3,149  + (3,000)
 10,264   (10,000)
```

4.
```
   11
  6,732    (7,000)
  3,152    (3,000)
+7,321  + (7,000)
 17,205   (17,000)
```

7. $2,365 + 1,295 + 3,116 = 6,776$ fish

8. $7,851 + 3,895 = 11,746$ students

```
       11
5.    5,232      (5,000)
      7,111      (7,000)
     +3,765    + (4,000)
     16,108     (16,000)
```

```
       11
6.    1,257      (1,000)
      6,463      (6,000)
     +8,765    + (9,000)
     16,485     (16,000)
```

7. $1,083 + 482 + 1,216 + 741 = 3,522$ miles

8. $3,187 + 408 + 1,443 + 1,837 = 6,875$ miles

Lesson Practice 19B

```
      111
1.    2,817      (3,000)
      9,236      (9,000)
     +3,680    + (4,000)
     15,733     (16,000)
```

```
       1
2.    5,740      (6,000)
      4,221      (4,000)
     +3,321    + (3,000)
     13,282     (13,000)
```

```
      111
3.    3,213      (3,000)
      1,357      (1,000)
     +2,798    + (3,000)
      7,368      (7,000)
```

```
      211
4.    1,476      (1,000)
        746      (1,000)
     +9,813    + (10,000)
     12,035     (12,000)
```

```
       11
5.    2,741      (3,000)
      4,374      (4,000)
     +3,354    + (3,000)
     10,469     (10,000)
```

```
      121
6.    4,123      (4,000)
      7,491      (7,000)
     +  486    + (0,000)
     12,100     (11,000)
```

Lesson Practice 19C

```
      121
1.    3,695      (4,000)
      3,175      (3,000)
     +2,141    + (2,000)
      9,011      (9,000)
```

```
       11
2.    1,468      (1,000)
      6,012      (6,000)
     +5,280    + (5,000)
     12,760     (12,000)
```

```
      211
3.    2,940      (3,000)
      4,278      (4,000)
     +1,963    + (2,000)
      9,181      (9,000)
```

```
      112
4.    1,086      (1,000)
      3,608      (4,000)
     +2,657    + (3,000)
      7,351      (8,000)
```

```
       11
5.    1,340      (1,000)
      1,765      (2,000)
     +8,041    + (8,000)
     11,146     (11,000)
```

```
      111
6.    4,729      (5,000)
      7,761      (8,000)
     +   35    + (0,000)
     12,525     (13,000)
```

7. $8,716 + 6,658 = 15,374$ people

8. $\$2,345 + \$5,634 + \$1,954 = \$9,933$

Systematic Review19D

1.
```
  121
  2,475
  1,890
+   376
  4,741
```

2.
```
     1
  7,513
  9,025
 +3,254
 19,792
```

3.
```
    111
  3,189
  1,422
 +2,468
  7,079
```

4. $8 - 4 = 4$
5. $10 - 6 = 4$
6. $14 - 7 = 7$
7. $10 - 9 = 1$
8. $6 - 3 = 3$
9. $10 - 3 = 7$
10. $4 - 2 = 2$
11. $10 - 5 = 5$
12. $6 > 4$
13. $14 > 6$
14. $3 < 7$
15. $\$35 + \$42 + \$33 + \$45 = \$155$
16. $\$17 - \$8 = \$9$
17. $3,645 + 4,782 + 5,641 = 14,068$ earthworms
18. $5 + 5 + 5 = 15"$

Systematic Review19E

1.
```
     11
  2,384
  4,123
 +6,335
 12,842
```

2.
```
    121
  3,591
  2,367
 +8,459
 14,417
```

3.
```
     11
  2,368
  4,152
 +6,314
 12,834
```

4. $9 - 5 = 4$
5. $2 - 1 = 1$
6. $9 - 6 = 3$
7. $10 - 2 = 8$
8. $9 - 3 = 6$
9. $16 - 9 = 7$
10. $9 - 4 = 5$
11. $9 - 7 = 2$
12. $9 < 15$
13. $9 = 9$
14. $2 < 8$
15. $\$2.98 + \$3.75 = \$6.73$
16. $7 - 5 = 2$ dimes

    2 dimes = 20¢
17. $20 + 18 + 10 + 35 = 83$ tons
18. $35 + 35 + 35 + 35 = 140'$

Systematic Review19F

1.
```
     11
  8,482
  4,621
 +5,351
 18,454
```

2.
```
     11
  1,264
  7,632
 +1,953
 10,849
```

3.
```
     11
  5,148
  2,633
 +4,186
 11,967
```

4. $7 - 3 = 4$
5. $11 - 6 = 5$
6. $7 - 4 = 3$
7. $13 - 5 = 8$

8. $8 - 3 = 5$

9. $9 - 2 = 7$

10. $8 - 5 = 3$

11. $11 - 4 = 7$

12. $2 > 1$

13. $5 < 19$

14. $8 = 8$

15. $495 + 382 + 516 + 402 = 1{,}795$ miles

16. $25 + 35 + 25 + 35 = 120'$

17. 2, 4, 6, 8, 10, 12 mittens

18. $12 - 4 = 8$ mittens

Lesson Practice 20A

1. done

2.
```
  60     40
 -40    +20
  20     60
```

3.
```
  94     51
 -51    +43
  43     94
```

4.
```
  53     42
 -42    +11
  11     53
```

5.
```
  40     30
 -30    +10
  10     40
```

6.
```
 459    312
-312   +147
 147    459
```

7.
```
  44     20
 -20    +24
  24     44
```

8.
```
 924     13
 -13   +911
 911    924
```

9.
```
 506    302
-302   +204
 204    506
```

10.
```
  25     21
 -21    + 4
   4     25
```

11.
```
 841    620
-620   +221
 221    841
```

12.
```
 999    123
-123   +876
 876    999
```

13. $32 - 10 = 22$ students

14. $44 - 11 = 33$ rabbits

Lesson Practice 20B

1.
```
  35     24
 -24    +11
  11     35
```

2.
```
  26     13
 -13    +13
  13     26
```

3.
```
  50     20
 -20    +30
  30     50
```

4.
```
  83     12
 -12    +71
  71     83
```

5.
```
  49     47
 -47    + 2
   2     49
```

6.
```
 989    432
-432   +557
 557    989
```

7.
```
  46     30
 -30    +16
  16     46
```

8.
```
 554     21
- 21   +533
 533    554
```

9.
```
 300    100
-100   +200
 200    300
```

10.
```
  62     30
 -30    +32
  32     62
```

11.
```
 438    214
-214   +224
 224    438
```

12.
```
 397    175
-175   +222
 222    397
```

13. $68 - 25 = 43$ cards

14. $48 - 23 = 25$ gallons

Lesson Practice 20C

1.
$$\begin{array}{r} 65 \\ -32 \\ \hline 33 \end{array} \quad \begin{array}{r} 32 \\ +33 \\ \hline 65 \end{array}$$

2.
$$\begin{array}{r} 17 \\ -17 \\ \hline 0 \end{array} \quad \begin{array}{r} 17 \\ + \ 0 \\ \hline 17 \end{array}$$

3.
$$\begin{array}{r} 52 \\ -21 \\ \hline 31 \end{array} \quad \begin{array}{r} 21 \\ +31 \\ \hline 52 \end{array}$$

4.
$$\begin{array}{r} 20 \\ -10 \\ \hline 10 \end{array} \quad \begin{array}{r} 10 \\ +10 \\ \hline 20 \end{array}$$

5.
$$\begin{array}{r} 75 \\ -32 \\ \hline 43 \end{array} \quad \begin{array}{r} 32 \\ +43 \\ \hline 75 \end{array}$$

6.
$$\begin{array}{r} 188 \\ - \ 23 \\ \hline 165 \end{array} \quad \begin{array}{r} 23 \\ +165 \\ \hline 188 \end{array}$$

7.
$$\begin{array}{r} 69 \\ -31 \\ \hline 38 \end{array} \quad \begin{array}{r} 31 \\ +38 \\ \hline 69 \end{array}$$

8.
$$\begin{array}{r} 561 \\ -260 \\ \hline 301 \end{array} \quad \begin{array}{r} 260 \\ +301 \\ \hline 561 \end{array}$$

9.
$$\begin{array}{r} 645 \\ -435 \\ \hline 210 \end{array} \quad \begin{array}{r} 435 \\ +210 \\ \hline 645 \end{array}$$

10.
$$\begin{array}{r} 85 \\ -44 \\ \hline 41 \end{array} \quad \begin{array}{r} 44 \\ +41 \\ \hline 85 \end{array}$$

11.
$$\begin{array}{r} 225 \\ -100 \\ \hline 125 \end{array} \quad \begin{array}{r} 100 \\ +125 \\ \hline 225 \end{array}$$

12.
$$\begin{array}{r} 538 \\ -421 \\ \hline 117 \end{array} \quad \begin{array}{r} 421 \\ +117 \\ \hline 538 \end{array}$$

13. $29 - 13 = 16$ children

14. $\$49 - \$21 = \$28$

2.
$$\begin{array}{r} 50 \\ -40 \\ \hline 10 \end{array} \quad \begin{array}{r} 40 \\ +10 \\ \hline 50 \end{array}$$

3.
$$\begin{array}{r} 39 \\ -27 \\ \hline 12 \end{array} \quad \begin{array}{r} 27 \\ +12 \\ \hline 39 \end{array}$$

4.
$$\begin{array}{r} 693 \\ -361 \\ \hline 332 \end{array} \quad \begin{array}{r} 361 \\ +332 \\ \hline 693 \end{array}$$

5.
$$\begin{array}{r} 300 \\ -100 \\ \hline 200 \end{array} \quad \begin{array}{r} 100 \\ +200 \\ \hline 300 \end{array}$$

6.
$$\begin{array}{r} 163 \\ - \ 51 \\ \hline 112 \end{array} \quad \begin{array}{r} 51 \\ +112 \\ \hline 163 \end{array}$$

7.
$$\begin{array}{r} \overset{111}{} \\ 2,174 \\ 7,418 \\ +3,791 \\ \hline 13,383 \end{array}$$

8.
$$\begin{array}{r} \overset{111}{} \\ 2,564 \\ 6,408 \\ +1,243 \\ \hline 10,215 \end{array}$$

9.
$$\begin{array}{r} \overset{12}{} \\ 379 \\ 511 \\ 333 \\ +468 \\ \hline 1,691 \end{array}$$

10. 124,971; "one hundred twenty-four thousand, nine hundred seventy-one"

11. 2, 4, 6, 8, 10, 12, 14, 16, 18, 20

12. 5, 10, 15, 20, 25, 30, 35, 40, 45, 50

13. 10, 20, 30, 40, 50, 60, 70, 80, 90, 100

14. $\$.25 - \$.14 = \$.11$

15. $103 + 3,521 + 58 = 3,682$ miles

Systematic Review 20D

1.
$$\begin{array}{r} 77 \\ -11 \\ \hline 66 \end{array} \quad \begin{array}{r} 11 \\ +66 \\ \hline 77 \end{array}$$

Systematic Review 20E

1.
$$\begin{array}{r} 35 \\ -22 \\ \hline 13 \end{array} \quad \begin{array}{r} 22 \\ +13 \\ \hline 35 \end{array}$$

2.
```
   49      31
 - 31    + 18
   18      49
```

3.
```
   28       5
 -  5    + 23
   23      28
```

4.
```
  633     510
 -510    + 123
  123     633
```

5.
```
  790     690
 -690    + 100
  100     790
```

6.
```
  561     550
 -550    +  11
   11     561
```

7.
```
      12
   1,890
   3,672
 + 7,254
  12,816
```

8.
```
     121
   7,193
   4,685
 + 1,492
  13,370
```

9.
```
      11
     922
     678
     253
   + 112
   1,965
```

10. 205,918; "two hundred five thousand, nine hundred eighteen"

11. 5, 10, 15, 20, 25, 30, 35, 40, 45, 50

12. 2, 4, 6, 8, 10, 12, 14, 16, 18, 20

13. 10, 20, 30, 40, 50, 60, 70, 80, 90, 100

14. $.49 − $.23 = $.26

15. 285 + 683 + 491 = 1,459 boxes

Systematic Review 20F

1.
```
   90      70
 - 70    + 20
   20      90
```

2.
```
   51      10
 - 10    + 41
   41      51
```

3.
```
   74      31
 - 31    + 43
   43      74
```

4.
```
  139     125
 -125    +  14
   14     139
```

5.
```
  999     347
 -347    + 652
  652     999
```

6.
```
  167      24
 - 24    + 143
  143     167
```

7.
```
     111
   2,467
   9,221
 + 6,788
  18,476
```

8.
```
      11
   4,253
   4,112
 + 1,255
   9,620
```

9.
```
      12
     238
     412
     633
   + 459
   1,742
```

10. 23,684; "twenty-three thousand, six hundred eighty-four"

11. 10, 20, 30, 40, 50, 60, 70, 80, 90, 100

12. 5, 10, 15, 20, 25, 30, 35, 40, 45, !

13. 2, 4, 6, 8, 10, 12, 14, 16, 18, 20

14. 12 + 12 + 12 = 36"

15. 67 − 13 = 54 cars

Lesson Practice 21A

1. done
2. : 35
3. : 40
4. : 10
5. : 48
6. : 47

Lesson Practice 21B

1. : 15
2. : 25
3. : 55
4. : 50
5. : 29
6. : 16

Lesson Practice 21C

1. : 05
2. : 55
3. : 40
4. : 00
5. : 08
6. : 32

Systematic Review 21D

1. : 10
2. : 54
3.
$$\begin{array}{r} 99 \\ -84 \\ \hline 15 \end{array} \qquad \begin{array}{r} 84 \\ +15 \\ \hline 99 \end{array}$$
4.
$$\begin{array}{r} 138 \\ -\ 34 \\ \hline 104 \end{array} \qquad \begin{array}{r} 34 \\ +104 \\ \hline 138 \end{array}$$
5.
$$\begin{array}{r} 279 \\ -164 \\ \hline 115 \end{array} \qquad \begin{array}{r} 164 \\ +115 \\ \hline 279 \end{array}$$
6.
$$\begin{array}{r} 1 \\ \$3.27 \\ +4.16 \\ \hline \$7.43 \end{array}$$

7.
$$\begin{array}{r} 1 \\ 2,384 \\ +6,335 \\ \hline 8,719 \end{array}$$
8.
$$\begin{array}{r} 11 \\ 367 \\ 591 \\ +419 \\ \hline 1,377 \end{array}$$
9. $4,568 + 3,851 + 2,001 = 10,420$ insects
10. $10 + 10 + 10 + 10 = 40'$ perimeter
    $40' - 3' = 37'$ of fence
11. $25 - 14 = 11$ years
12. $19 - 7 = 12$ points

Systematic Review 21E

1. : 20
2. : 49
3.
$$\begin{array}{r} 83 \\ -52 \\ \hline 31 \end{array} \qquad \begin{array}{r} 52 \\ +31 \\ \hline 83 \end{array}$$
4.
$$\begin{array}{r} 873 \\ -\ 61 \\ \hline 812 \end{array} \qquad \begin{array}{r} 61 \\ +812 \\ \hline 873 \end{array}$$
5.
$$\begin{array}{r} 362 \\ -151 \\ \hline 211 \end{array} \qquad \begin{array}{r} 151 \\ +211 \\ \hline 362 \end{array}$$
6.
$$\begin{array}{r} 1 \\ \$\ 9.28 \\ +\ 2.17 \\ \hline \$11.45 \end{array}$$
7.
$$\begin{array}{r} 4,123 \\ +7,460 \\ \hline 11,583 \end{array}$$
8.
$$\begin{array}{r} 21 \\ 45\text{ə} \\ 3\text{є.} \\ +591 \\ \hline 1,417 \end{array}$$
9. $32 + 28 = 60$ years
10. $29 - 8 = 21$ years
11. $40 + 35 + 40 + 35 = 150'$
12. $144 + 120 = 264$ made
    $264 - 52 = 212$ eaten

Systematic Review 21F

1. $:40$

2. $:17$

3. $\begin{array}{r} 19 \\ -7 \\ \hline 12 \end{array}$ $\begin{array}{r} 7 \\ +12 \\ \hline 19 \end{array}$

4. $\begin{array}{r} 839 \\ -16 \\ \hline 823 \end{array}$ $\begin{array}{r} 16 \\ +823 \\ \hline 839 \end{array}$

5. $\begin{array}{r} 604 \\ -302 \\ \hline 302 \end{array}$ $\begin{array}{r} 302 \\ +302 \\ \hline 604 \end{array}$

6. $\begin{array}{r} \$5.36 \\ +1.5\,1 \\ \hline \$6.87 \end{array}$

7. $\begin{array}{r} {}^{11} \\ 8,482 \\ +4,621 \\ \hline 13,103 \end{array}$

8. $\begin{array}{r} {}^{11} \\ 343 \\ 246 \\ +112 \\ \hline 701 \end{array}$

9. 4 dimes = 40¢

   3 nickels = 15¢

   $\$.40 + \$.15 = \$.55$

   $\$.55 - \$.25 = \$.30$

10. $6 + 6 + 6 = 18''$

11. $969 + 1{,}345 + 5{,}002 + 2{,}061 = 9{,}377$ leaves

12. $\$25 - \$10 = \$15$

Lesson Practice 22A

1. done

2. $\begin{array}{r} {}^{5} \\ \cancel{6}\,{}^{1}1 \\ -2\,7 \\ \hline 3\,4 \end{array}$ $\begin{array}{r} 1 \\ 27 \\ +34 \\ \hline 6\,1 \end{array}$

3. $\begin{array}{r} {}^{1} \\ \cancel{2}\,{}^{1}2 \\ -1\,3 \\ \hline 9 \end{array}$ $\begin{array}{r} 1 \\ 13 \\ +9 \\ \hline 22 \end{array}$

4. $\begin{array}{r} {}^{11} \\ \cancel{2}\,{}^{1}3 \\ -1\,9 \\ \hline 4 \end{array}$ $\begin{array}{r} 19 \\ +4 \\ \hline 23 \end{array}$

5. $\begin{array}{r} {}^{4} \\ \cancel{5}\,{}^{1}5 \\ -2\,6 \\ \hline 2\,9 \end{array}$ $\begin{array}{r} 1 \\ 26 \\ +29 \\ \hline 55 \end{array}$

6. $\begin{array}{r} {}^{2} \\ \cancel{3}\,{}^{1}2 \\ -1\,4 \\ \hline 1\,8 \end{array}$ $\begin{array}{r} 1 \\ 14 \\ +18 \\ \hline 32 \end{array}$

7. $\begin{array}{r} {}^{6} \\ \cancel{7}\,{}^{1}3 \\ -3\,4 \\ \hline 3\,9 \end{array}$ $\begin{array}{r} 1 \\ 34 \\ +39 \\ \hline 73 \end{array}$

8. $\begin{array}{r} {}^{5} \\ \cancel{6}\,{}^{1}8 \\ -2\,9 \\ \hline 3\,9 \end{array}$ $\begin{array}{r} 1 \\ 29 \\ +39 \\ \hline 68 \end{array}$

9. $\begin{array}{r} {}^{5} \\ \cancel{6}\,{}^{1}3 \\ -4\,9 \\ \hline 1\,4 \end{array}$ $\begin{array}{r} 1 \\ 49 \\ +14 \\ \hline 63 \end{array}$

10. $\$.75 - \$.57 = \$.18$

11. $43 - 28 = 15$ barrettes

12. $72 - 65 = 7$ peanuts

Lesson Practice 22B

1. $\begin{array}{r} {}^{4} \\ \cancel{5}\,{}^{1}7 \\ -2\,9 \\ \hline 2\,8 \end{array}$ $\begin{array}{r} 1 \\ 29 \\ +28 \\ \hline 57 \end{array}$

2. $\begin{array}{r} {}^{2} \\ \cancel{3}\,{}^{1}0 \\ -1\,8 \\ \hline 1\,2 \end{array}$ $\begin{array}{r} 1 \\ 18 \\ +12 \\ \hline 30 \end{array}$

3. $\begin{array}{r} {}^{5} \\ \cancel{6}\,{}^{1}5 \\ -4\,7 \\ \hline 1\,8 \end{array}$ $\begin{array}{r} 1 \\ 47 \\ +18 \\ \hline 65 \end{array}$

4. $\begin{array}{r} {}^{4} \\ \cancel{5}\,{}^{1}2 \\ -1\,4 \\ \hline 3\,8 \end{array}$ $\begin{array}{r} 1 \\ 14 \\ +38 \\ \hline 52 \end{array}$

5. $\cancel{8}^{7}{}^{1}5$ → 49
    -49 $+36$
    ─────   ─────
    36 85

6. $\cancel{7}^{6}{}^{1}2$ → 27
    -27 $+45$
    ─────   ─────
    45 72

7. $\cancel{5}^{4}{}^{1}3$ → 18
    -18 $+35$
    ─────   ─────
    35 53

8. $\cancel{3}^{2}{}^{1}1$ → 16
    -16 $+15$
    ─────   ─────
    15 31

9. $\cancel{8}^{7}{}^{1}2$ → 53
    -53 $+29$
    ─────   ─────
    29 82

10. $43 - 24 = 19$ pages
11. $61 - 36 = 25$ times
12. $\$.50 - \$.23 = \$.27$

Lesson Practice 22C

1. $\cancel{5}^{4}{}^{1}2$ → 25
    -25 $+27$
    ─────   ─────
    27 52

2. $\cancel{7}^{6}{}^{1}1$ → 3
    $-\ 3$ $+68$
    ─────   ─────
    68 71

3. $\cancel{3}^{2}{}^{1}4$ → 15
    -15 $+19$
    ─────   ─────
    19 34

4. $\cancel{8}^{7}{}^{1}7$ → 8
    $-\ 8$ $+79$
    ─────   ─────
    79 87

5. $\cancel{6}^{5}{}^{1}2$ → 27
    -27 $+35$
    ─────   ─────
    35 62

6. $\cancel{2}^{1}{}^{1}3$ → 14
    -14 $+\ 9$
    ─────   ─────
    9 23

7. $\cancel{4}^{3}{}^{1}5$ → 26
    -26 $+19$
    ─────   ─────
    19 45

8. $\cancel{3}^{2}{}^{1}1$ → 29
    -29 $+\ 2$
    ─────   ─────
    2 31

9. $\cancel{7}^{6}{}^{1}2$ → 38
    -38 $+34$
    ─────   ─────
    34 72

10. $65 - 6 = 59$ pennies
11. $48 - 29 = 19$ years
12. $63 - 24 = 39$ apples

Systematic Review 22D

1. $\cancel{7}^{6}{}^{1}1$ → 43
    -43 $+28$
    ─────   ─────
    28 71

2. $\cancel{9}^{8}{}^{1}8$ → 89
    -89 $+\ 9$
    ─────   ─────
    9 98

3. $\cancel{3}^{2}{}^{1}5$ → 17
    -17 $+18$
    ─────   ─────
    18 35

4. 85 45
    -45 $+40$
    ─────   ─────
    40 85

5. 429 311
    -311 $+118$
    ─────   ─────
    118 429

6. 148 26
    $-\ 26$ $+122$
    ─────   ─────
    122 148

7.
$$\begin{array}{r} 1 \\ \$7.33 \\ +1.83 \\ \hline \$9.16 \end{array}$$

8.
$$\begin{array}{r} 1 \\ 5,263 \\ +7,554 \\ \hline 12,817 \end{array}$$

9.
$$\begin{array}{r} 11 \\ 892 \\ 413 \\ +476 \\ \hline 1,781 \end{array}$$

10. : 13

11. : 45

12. $16 + 21 + 14 = 51$
$51 - 25 = 26$ marbles

8.
$$\begin{array}{r} 9,435 \\ +4,252 \\ \hline 13,687 \end{array}$$

9.
$$\begin{array}{r} 21 \\ 573 \\ 882 \\ 235 \\ +762 \\ \hline 2,452 \end{array}$$

10. : 30

11. : 53

12. $19 + 21 = 40$
$52 - 40 = 12$ pages

Systematic Review 22E

1.
$$\begin{array}{cc} 8 & 1 \\ 9\,{}^{1}2 & 78 \\ -78 & +14 \\ \hline 14 & 92 \end{array}$$

2.
$$\begin{array}{cc} 4 & 1 \\ 5\,{}^{1}3 & 45 \\ -45 & +\,8 \\ \hline 8 & 53 \end{array}$$

3.
$$\begin{array}{cc} 1 & 1 \\ 2\,{}^{1}3 & 14 \\ -14 & +\,9 \\ \hline 9 & 23 \end{array}$$

4.
$$\begin{array}{cc} 75 & 50 \\ -50 & +25 \\ \hline 25 & 75 \end{array}$$

5.
$$\begin{array}{cc} 743 & 12 \\ -\,12 & +731 \\ \hline 731 & 743 \end{array}$$

6.
$$\begin{array}{cc} 514 & 512 \\ -\,2 & +\,2 \\ \hline 512 & 514 \end{array}$$

7.
$$\begin{array}{r} 1 \\ \$3.55 \\ +2.52 \\ \hline \$6.07 \end{array}$$

Systematic Review 22F

1.
$$\begin{array}{cc} 3 & 1 \\ 4\,{}^{1}4 & 28 \\ -28 & +16 \\ \hline 16 & 44 \end{array}$$

2.
$$\begin{array}{cc} 8 & 1 \\ 9\,{}^{1}3 & 16 \\ -16 & +77 \\ \hline 77 & 93 \end{array}$$

3.
$$\begin{array}{cc} 7 & 1 \\ 8\,{}^{1}4 & 27 \\ -27 & +57 \\ \hline 57 & 84 \end{array}$$

4.
$$\begin{array}{cc} 17 & 10 \\ -10 & +\,7 \\ \hline 7 & 17 \end{array}$$

5.
$$\begin{array}{cc} 913 & 112 \\ -112 & +801 \\ \hline 801 & 913 \end{array}$$

6.
$$\begin{array}{cc} 307 & 205 \\ -205 & +102 \\ \hline 102 & 307 \end{array}$$

7.
$$\begin{array}{r} {}^{1} \\ \$\,6.24 \\ +\;5.48 \\ \hline \$11.72 \end{array}$$

8.
$$\begin{array}{r} 111 \\ 3,548 \\ 7,624 \\ +3,642 \\ \hline 14,814 \end{array}$$

9.
$$
\begin{array}{r}
2\\
573\\
882\\
+762\\
\hline
2{,}217
\end{array}
$$

10. :02

11. :15

12. Bill: $134+48+2=184$ animals
Tom: $100+22=122$ animals
$184-122=62$ more

Lesson Practice 23A

1. done
2. 7 : 00
3. 11 : 00
4. 1 : 00

Lesson Practice 23B

1. 8 : 00
2. 2 : 00
3. 12 : 00
4. 4 : 00

Lesson Practice 23C

1. done
2. 5 : 45
3. 3 : 40
4. 12 : 25

Systematic Review 23D

1. 5 : 50
2. 2 : 40
3. 7 : 09
4. 4 : 17

5.
$$
\begin{array}{rr}
\overset{3}{\cancel{4}}{}^{1}0 & \overset{1}{38}\\
-38 & +2\\
\hline
2 & 40
\end{array}
$$

6.
$$
\begin{array}{rr}
\overset{6}{\cancel{7}}{}^{1}6 & \overset{1}{18}\\
-18 & +58\\
\hline
58 & 76
\end{array}
$$

7.
$$
\begin{array}{rr}
\overset{5}{\cancel{6}}{}^{1}2 & \overset{1}{57}\\
-57 & +5\\
\hline
5 & 62
\end{array}
$$

8.
$$
\begin{array}{rr}
\overset{6}{\cancel{7}}{}^{1}3 & \overset{1}{14}\\
-14 & +59\\
\hline
59 & 73
\end{array}
$$

9. $\$.40+\$.35+\$.08=\$.83$

10. $1{,}342+1.342+1{,}342+1{,}342=5{,}368'$

Systematic Review 23E

1. 6 : 00
2. 8 : 05
3. 1 : 21
4. 9 : 57

5.
$$
\begin{array}{rr}
\overset{6}{\cancel{7}}{}^{1}3 & \overset{1}{15}\\
-15 & +58\\
\hline
58 & 73
\end{array}
$$

6.
$$
\begin{array}{rr}
88 & 44\\
-44 & +44\\
\hline
44 & 88
\end{array}
$$

7.
$$
\begin{array}{rr}
\overset{6}{\cancel{7}}{}^{1}6 & \overset{1}{28}\\
-28 & +48\\
\hline
48 & 76
\end{array}
$$

8.
$$
\begin{array}{rr}
928 & 605\\
-605 & +323\\
\hline
323 & 928
\end{array}
$$

9. $12+12+12+12+12+12=72"$

10. $250+75=325$ soldiers

Systematic Review 23F

1. 10:15
2. 3:30
3. 12:08
4. 7:45

5.
$$\begin{array}{r} 5\,{}^{4}\!\!\!\!4 \\ -45 \\ \hline 9 \end{array} \quad \begin{array}{r} 1 \\ 45 \\ +\ 9 \\ \hline 54 \end{array}$$

6.
$$\begin{array}{r} 39 \\ -20 \\ \hline 19 \end{array} \quad \begin{array}{r} 20 \\ +19 \\ \hline 39 \end{array}$$

7.
$$\begin{array}{r} {}^{3}\!\!\!\!4\,{}^{1}0 \\ -35 \\ \hline 5 \end{array} \quad \begin{array}{r} 1 \\ 35 \\ +\ 5 \\ \hline 40 \end{array}$$

8.
$$\begin{array}{r} 779 \\ -314 \\ \hline 465 \end{array} \quad \begin{array}{r} 314 \\ +465 \\ \hline 779 \end{array}$$

9. $4.52 + $2.91 = $7.43
10. 35 − 26 = 9 birds

Lesson Practice 24A

1. done
2. done

3.
$$\begin{array}{r} {}^{1}\!\!\!\!2\,{}^{1}\!\!\!\!2\,{}^{1}3 \\ -\ 87 \\ \hline 136 \end{array} \quad \begin{array}{r} 11 \\ 87 \\ +136 \\ \hline 223 \end{array}$$

4.
$$\begin{array}{r} {}^{6}\!\!\!\!7\,{}^{2}\!\!\!\!3\,{}^{1}4 \\ -\ 36 \\ \hline 698 \end{array} \quad \begin{array}{r} 11 \\ 36 \\ +698 \\ \hline 734 \end{array}$$

5.
$$\begin{array}{r} {}^{4}\!\!\!\!5\,{}^{1}\!\!\!\!2\,{}^{1}3 \\ -138 \\ \hline 385 \end{array} \quad \begin{array}{r} 11 \\ 138 \\ +385 \\ \hline 523 \end{array}$$

6.
$$\begin{array}{r} {}^{3}\!\!\!\!4\,{}^{9}\!\!\!\!0\,{}^{1}0 \\ -399 \\ \hline 1 \end{array} \quad \begin{array}{r} 11 \\ 399 \\ +\ 1 \\ \hline 400 \end{array}$$

7.
$$\begin{array}{r} {}^{2}\!\!\!\!3\,{}^{1}\!\!\!\!2\,{}^{1}1 \\ -\ 39 \\ \hline 282 \end{array} \quad \begin{array}{r} 11 \\ 39 \\ +282 \\ \hline 321 \end{array}$$

8.
$$\begin{array}{r} 459 \\ -241 \\ \hline 218 \end{array} \quad \begin{array}{r} 241 \\ +218 \\ \hline 459 \end{array}$$

9.
$$\begin{array}{r} {}^{5}\!\!\!\!6\,{}^{1}0\,{}^{1}4 \\ -196 \\ \hline 418 \end{array} \quad \begin{array}{r} 11 \\ 196 \\ +418 \\ \hline 614 \end{array}$$

10. 235 − 167 = 68 pages
11. 300 − 245 = 55 pennies
12. 580 − 425 = 155'

Lesson Practice 24B

1.
$$\begin{array}{r} {}^{6}\!\!\!\!7\,{}^{0}\!\!\!\!1\,{}^{1}6 \\ -\ 58 \\ \hline 658 \end{array} \quad \begin{array}{r} 11 \\ 58 \\ +658 \\ \hline 716 \end{array}$$

2.
$$\begin{array}{r} {}^{5}\!\!\!\!6\,{}^{1}68 \\ -284 \\ \hline 384 \end{array} \quad \begin{array}{r} 1 \\ 284 \\ +384 \\ \hline 668 \end{array}$$

3.
$$\begin{array}{r} 163 \\ -\ 72 \\ \hline 91 \end{array} \quad \begin{array}{r} 1 \\ 72 \\ +\ 91 \\ \hline 163 \end{array}$$

4.
$$\begin{array}{r} {}^{5}\!\!\!\!6\,{}^{3}\!\!\!\!4\,{}^{1}1 \\ -\ 49 \\ \hline 592 \end{array} \quad \begin{array}{r} 11 \\ 49 \\ +592 \\ \hline 641 \end{array}$$

5.
$$\begin{array}{r} {}^{6}\!\!\!\!\,3\,7\,{}^{1}2 \\ -106 \\ \hline 266 \end{array} \quad \begin{array}{r} 1 \\ 106 \\ +266 \\ \hline 372 \end{array}$$

6.
$$\begin{array}{r} {}^{8}\!\!\!\!9\,{}^{7}\!\!\!\!8\,{}^{1}7 \\ -789 \\ \hline 198 \end{array} \quad \begin{array}{r} 11 \\ 789 \\ +198 \\ \hline 987 \end{array}$$

7.
$$\begin{array}{r} {}^{6}\ \\ \cancel{7}\,{}^{1}05 \\ -\ \ 60 \\ \hline 645 \end{array}$$
$$\begin{array}{r} 1 \\ 60 \\ +645 \\ \hline 705 \end{array}$$

8.
$$\begin{array}{r} {}^{4}\ \\ \cancel{5}\,{}^{1}00 \\ -250 \\ \hline 250 \end{array}$$
$$\begin{array}{r} 1 \\ 250 \\ +250 \\ \hline 500 \end{array}$$

9.
$$\begin{array}{r} {}^{6}\ \\ 4\cancel{7}\,{}^{1}1 \\ -106 \\ \hline 365 \end{array}$$
$$\begin{array}{r} 1 \\ 106 \\ +365 \\ \hline 471 \end{array}$$

10. $275 - $116 = $159

11. 553 - 378 = 175 acorns

12. 325 - 185 = 140 miles

7.
$$\begin{array}{r} {}^{9}\ \\ \cancel{1}\cancel{0}\,{}^{1}3 \\ -\ 25 \\ \hline 78 \end{array}$$
$$\begin{array}{r} 11 \\ 25 \\ +78 \\ \hline 103 \end{array}$$

8.
$$\begin{array}{r} {}^{4}\ \\ \cancel{5}\,{}^{1}72 \\ -390 \\ \hline 182 \end{array}$$
$$\begin{array}{r} 1 \\ 390 \\ +182 \\ \hline 572 \end{array}$$

9.
$$\begin{array}{r} {}^{2}\,{}^{1}2 \\ \cancel{3}\cancel{3}\,{}^{1}3 \\ -144 \\ \hline 189 \end{array}$$
$$\begin{array}{r} 11 \\ 144 \\ +189 \\ \hline 333 \end{array}$$

10. 875 - 80 = 795 people

11. 200 - 115 = 85 cards

12. 375 - 18 = 357 yards

Lesson Practice 24C

1.
$$\begin{array}{r} {}^{7}\ \\ 5\cancel{8}\,{}^{1}3 \\ -\ 64 \\ \hline 519 \end{array}$$
$$\begin{array}{r} 1 \\ 64 \\ +519 \\ \hline 583 \end{array}$$

2.
$$\begin{array}{r} {}^{1}\ \\ \cancel{2}\,{}^{1}03 \\ -192 \\ \hline 11 \end{array}$$
$$\begin{array}{r} 1 \\ 192 \\ +\ 11 \\ \hline 203 \end{array}$$

3.
$$\begin{array}{r} {}^{1}\,{}^{9} \\ \cancel{2}\cancel{0}\,{}^{1}0 \\ -\ 98 \\ \hline 102 \end{array}$$
$$\begin{array}{r} 11 \\ 98 \\ +102 \\ \hline 200 \end{array}$$

4.
$$\begin{array}{r} {}^{2}\ \\ \cancel{3}\,{}^{1}19 \\ -\ 30 \\ \hline 289 \end{array}$$
$$\begin{array}{r} 1 \\ 30 \\ +289 \\ \hline 319 \end{array}$$

5.
$$\begin{array}{r} {}^{2}\ \\ 6\cancel{3}\,{}^{1}1 \\ -429 \\ \hline 202 \end{array}$$
$$\begin{array}{r} 1 \\ 429 \\ +202 \\ \hline 631 \end{array}$$

6.
$$\begin{array}{r} {}^{3}\ \\ \cancel{4}\,{}^{1}19 \\ -138 \\ \hline 281 \end{array}$$
$$\begin{array}{r} 1 \\ 138 \\ +281 \\ \hline 419 \end{array}$$

Systematic Review 24D

1.
$$\begin{array}{r} {}^{9}\ \\ \cancel{1}\cancel{0}\,{}^{1}0 \\ -\ 75 \\ \hline 25 \end{array}$$
$$\begin{array}{r} 11 \\ 75 \\ +\ 25 \\ \hline 100 \end{array}$$

2.
$$\begin{array}{r} {}^{8}\ \\ \cancel{9}\,{}^{1}08 \\ -291 \\ \hline 617 \end{array}$$
$$\begin{array}{r} 1 \\ 291 \\ +617 \\ \hline 908 \end{array}$$

3.
$$\begin{array}{r} {}^{4}\ \\ 2\cancel{5}\,{}^{1}6 \\ -138 \\ \hline 118 \end{array}$$
$$\begin{array}{r} 1 \\ 138 \\ +118 \\ \hline 256 \end{array}$$

4.
$$\begin{array}{r} 199 \\ -\ 82 \\ \hline 117 \end{array}$$
$$\begin{array}{r} 82 \\ +117 \\ \hline 199 \end{array}$$

5.
$$\begin{array}{r} {}^{5}\ \\ \cancel{6}\,{}^{1}3 \\ -\ 24 \\ \hline 39 \end{array}$$
$$\begin{array}{r} 1 \\ 24 \\ +39 \\ \hline 63 \end{array}$$

6.
$$\begin{array}{r} {}^{3}\ \\ \cancel{4}\,{}^{1}1 \\ -\ 35 \\ \hline 6 \end{array}$$
$$\begin{array}{r} 1 \\ 35 \\ +\ 6 \\ \hline 41 \end{array}$$

7. 4:35

8. 11:12

9.
```
    436
  + 122
    558
```

10.
```
     11
    298
  + 539
    837
```

11.
```
    1 1 1
   2,999
  +3,111
   6,110
```

12. $175 - 98 = 77$ fireflies

13. $212 + 362 = 574$ miles

14. $215 + $134 = 349

$400 - $349 = 51

9.
```
   2 12
  + 362
    574
```

10.
```
     1
    356
  + 48 1
    837
```

11.
```
      1
   1,276
  +7,391
   8,667
```

12. $321 - 215 = 106$ pages

13. $665 + 133 = 798$ animals

14. $314 + 219 = 533$

$658 - 533 = 125$ cones

Systematic Review 24E

1.
```
   4
  15 12    87
  - 8 7  + 65
    6 5    152
            11
```

2.
```
    1
  3 2 10   1 18
  - 1 18  +202
   20 2    320
             1
```

3.
```
    803    300
   -300   +503
    503    803
```

4.
```
    8
  19 13    67
  - 6 7  +126
   12 6    193
             1
```

5.
```
    3
   4 13    17
  - 1 7  + 26
    2 6    43
             1
```

6.
```
    5
   6 14    38
  - 3 8  + 26
    2 6    64
             1
```

7. $2:05$

8. $4:48$

Systematic Review 24F

1.
```
    9
   10 10    47
  -  4 7  + 53
     5 3    100
             11
```

2.
```
    3
  1 4 11   1 13
  - 1 13  + 28
     2 8    141
             1
```

3.
```
   1 1
  2 2 10   164
  - 1 6 4  + 56
      5 6   220
            11
```

4.
```
    4
  15 10    98
  - 9 8  + 52
    5 2    150
            1
```

5.
```
    70    20
   -20   +50
    50    70
```

6.
```
    2
   3 12    18
  - 1 8  + 14
    1 4    32
            1
```

7. $5:22$

8. $10:36$

9.
```
    400
  + 376
    776
```

10.
$$\begin{array}{r} 11 \\ 288 \\ + 137 \\ \hline 425 \end{array}$$

11.
$$\begin{array}{r} 1 \\ 2,146 \\ + 1,408 \\ \hline 3,554 \end{array}$$

12. $581 - 499 = 82$ people

13. $\$126 + \$132 = \$258$

14. $45 + 11 + 5 + 8 = 69$
$100 - 69 = 31$ dogs

Lesson Practice 25A

1. February
2. second
3. first — Tuesday
4. second — Sunday
5. third — Friday
6. fourth — Thursday
7. fifth — Wednesday
8. sixth — Saturday
9. seventh — Monday
10. 31
11. 31
12. 30
13. 5
14. 3
15. 8
16. ||||| ||||| ||||| ||||
17. ||||| ||||| ||||| ||||| ||||| |||||
18. ||||| ||||| ||||| ||||| ||||| ||||
19. ||||| ||||| ||||| ||||| ||||| ||||| |||||
20. sixth

Slash lines on tally marks
may slant either direction.

Lesson Practice 25B

1. Wednesday
2. seventh
3. October
4. first — April
5. second — February
6. third — June
7. fourth — January
8. fifth — March
9. sixth — May
10. 28 or 29
11. 31
12. 30
13. 11
14. 6
15. 15
16. ||||| |||||
17. ||||| ||||| |||
18. ||||| ||
19. 17 cars
20. Sunday

Lesson Practice 25C

1. April
2. third
3. first
4. seventh — September
5. eighth — December
6. ninth — July
7. tenth — November
8. eleventh — October
9. twelfth — August
10. 30
11. 31
12. 31
13. 12
14. 8
15. 14
16. ||||| ||||| ||||| ||||| |||||
17. ||||| ||||

18. | | | |
19. Thursday
20. Christmas

Systematic Review 25D

1. January
2. fourth
3. sixth
4. 31

5.
$$\begin{array}{r} 4 \\ 3\,\not5\,{}^1 5 \\ -\ \ 2\,6 \\ \hline 3\,2\,9 \end{array}$$

6.
$$\begin{array}{r} 5 \\ \not6\,{}^1 18 \\ -\ 2\,24 \\ \hline 3\,94 \end{array}$$

7.
$$\begin{array}{r} 3 \\ \not4\,{}^1 19 \\ -\ 3\,57 \\ \hline 62 \end{array}$$

8.
$$\begin{array}{r} 1\ \ 1 \\ 3{,}629 \\ +2{,}428 \\ \hline 6{,}057 \end{array}$$

9. 7 : 15
10. 4 : 55
11. 12 + 8 = 20 times
12. 5 + 17 + 14 = 36 beans

|||| |||| |||| |||| |||| |||| |||| |

13. 131 − 47 = 84 days
14. 11 + 11 + 11 + 11 = 44'

Systematic Review 25E

1. March
2. Friday
3. September
4. 30

5.
$$\begin{array}{r} 1 \\ \not2\,{}^1 34 \\ -\ \ 52 \\ \hline 1\,8\,2 \end{array}$$

6.
$$\begin{array}{r} 8 \\ 5\,\not9\,{}^1 0 \\ -4\,1\,8 \\ \hline 1\,7\,2 \end{array}$$

7.
$$\begin{array}{r} 8\ {}^1 0 \\ \not9\,\not1\,{}^1 0 \\ -\ \ 7\,5 \\ \hline 8\,3\,5 \end{array}$$

8.
$$\begin{array}{r} 1\ \ 1 \\ 1{,}831 \\ +5{,}439 \\ \hline 7{,}270 \end{array}$$

9. 3 : 00
10. 6 : 30
11. 12 times
12. answers will vary
13. 3,451 + 2,163 + 1,999 = 7,613 miles
14. August

Systematic Review 25F

1. May
2. November
3. sixth
4. 31

5.
$$\begin{array}{r} 4 \\ \not5\,{}^1 07 \\ -\ \ 4\,1 \\ \hline 4\,66 \end{array}$$

6.
$$\begin{array}{r} 7 \\ 1\,\not8\,{}^1 2 \\ -1\,4\,6 \\ \hline 3\,6 \end{array}$$

7.
$$\begin{array}{r} 8\,9 \\ \not9\,\not0\,{}^1 0 \\ -\ \ 2\,5 \\ \hline 8\,7\,5 \end{array}$$

8.
$$\begin{array}{r} 1 \\ 7{,}147 \\ +8{,}713 \\ \hline 15{,}860 \end{array}$$

9. $5:22$

10. $10:05$

11. $19 < 21$

David read more

12. seventh

13. $110 - 35 = 75$ years

14. $6 + 10 + 13 = 29"$

Lesson Practice 26A

1. done

2. done

3.
$$
\begin{array}{r} {}^{2}\,{}^{1}2 \\ 5,3\,3\,{}^{1}3 \\ -\,1,1\,8\,6 \\ \hline 4,1\,4\,7 \end{array}
\qquad
\begin{array}{r} 1\,1 \\ 1,1\,8\,6 \\ +\,4,1\,4\,7 \\ \hline 5,3\,3\,3 \end{array}
$$

4.
$$
\begin{array}{r} {}^{3}\ \ {}^{7} \\ 4,{}^{1}2\,8\,{}^{1}4 \\ -\ \ \ 9\,5\,5 \\ \hline 3,3\,2\,9 \end{array}
\qquad
\begin{array}{r} 1\ \ 1 \\ 9\,5\,5 \\ +\,3,3\,2\,9 \\ \hline 4,2\,8\,4 \end{array}
$$

5.
$$
\begin{array}{r} {}^{5} \\ 8,2\,6\,{}^{1}3 \\ -\,3,1\,4\,9 \\ \hline 5,1\,1\,4 \end{array}
\qquad
\begin{array}{r} 1 \\ 3,1\,4\,9 \\ +\,5,1\,1\,4 \\ \hline 8,2\,6\,3 \end{array}
$$

6.
$$
\begin{array}{r} {}^{2}\,{}^{1}1 \\ 3,2\,{}^{1}6\,4 \\ -\,2,5\,8\,2 \\ \hline 6\,8\,2 \end{array}
\qquad
\begin{array}{r} 1\,1 \\ 6\,8\,2 \\ +\,2,5\,8\,2 \\ \hline 3,2\,6\,4 \end{array}
$$

7.
$$
\begin{array}{r} {}^{6}\,{}^{1}\,{}^{1}3 \\ 7,2\,4\,{}^{1}1 \\ -\ \ \ 3\,7\,8 \\ \hline 6,8\,6\,3 \end{array}
\qquad
\begin{array}{r} 1\,1\,1 \\ 3\,7\,8 \\ +\,6,8\,6\,3 \\ \hline 7,2\,4\,1 \end{array}
$$

8.
$$
\begin{array}{r} {}^{5}\,{}^{9}\,{}^{9} \\ 6,0\,0\,{}^{1}0 \\ -\,5,1\,3\,9 \\ \hline 8\,6\,1 \end{array}
\qquad
\begin{array}{r} 1\,1\,1 \\ 5,1\,3\,9 \\ +\ \ \ 8\,6\,1 \\ \hline 6,0\,0\,0 \end{array}
$$

9.
$$
\begin{array}{r} {}^{6} \\ 6,7\,{}^{1}3\,2 \\ -\,3,1\,5\,2 \\ \hline 3,5\,8\,0 \end{array}
\qquad
\begin{array}{r} 1 \\ 3,1\,5\,2 \\ +\,3,5\,8\,0 \\ \hline 6,7\,3\,2 \end{array}
$$

10. $1,465 - 906 = 559$ ants

11. $2,375 - 1,490 = 885$ miles

12. $1,760 - 1,588 = 172$ years

Lesson Practice 26B

1.
$$
\begin{array}{r} {}^{4} \\ 5,{}^{1}0\,8\,9 \\ -\ \ \ 6\,3\,2 \\ \hline 4,4\,5\,7 \end{array}
\qquad
\begin{array}{r} 1 \\ 4,4\,5\,7 \\ +\ \ \ 6\,3\,2 \\ \hline 5,0\,8\,9 \end{array}
$$

2.
$$
\begin{array}{r} {}^{6}\,{}^{1} \\ 7,{}^{1}3\,2\,{}^{1}1 \\ -\,2,5\,1\,4 \\ \hline 4,8\,0\,7 \end{array}
\qquad
\begin{array}{r} 1\ \ 1 \\ 2,5\,1\,4 \\ +\,4,8\,0\,7 \\ \hline 7,3\,2\,1 \end{array}
$$

3.
$$
\begin{array}{r} {}^{8}\,{}^{9}\,{}^{9} \\ 9,0\,0\,{}^{1}0 \\ -\,1,2\,8\,7 \\ \hline 7,7\,1\,3 \end{array}
\qquad
\begin{array}{r} 1\,1\,1 \\ 1,2\,8\,7 \\ +\,7,7\,1\,3 \\ \hline 9,0\,0\,0 \end{array}
$$

4.
$$
\begin{array}{r} {}^{6}\,{}^{1}0\,{}^{1}0 \\ 7,1\,1\,{}^{1}1 \\ -\ \ \ 2\,3\,2 \\ \hline 6,8\,7\,9 \end{array}
\qquad
\begin{array}{r} 1\,1\,1 \\ 2\,3\,2 \\ +\,6,8\,7\,9 \\ \hline 7,1\,1\,1 \end{array}
$$

5.
$$
\begin{array}{r} {}^{4}\,{}^{1}2\,{}^{1}5 \\ 5,3\,6\,{}^{1}1 \\ -\,3,7\,6\,5 \\ \hline 1,5\,9\,6 \end{array}
\qquad
\begin{array}{r} 1\,1\,1 \\ 3,7\,6\,5 \\ +\,1,5\,9\,6 \\ \hline 5,3\,6\,1 \end{array}
$$

6.
$$
\begin{array}{r} {}^{0} \\ 7,2\,1\,{}^{1}4 \\ -\,1,1\,0\,8 \\ \hline 6,1\,0\,6 \end{array}
\qquad
\begin{array}{r} 1 \\ 1,1\,0\,8 \\ +\,6,1\,0\,6 \\ \hline 7,2\,1\,4 \end{array}
$$

7.
$$
\begin{array}{r} {}^{3}\,{}^{9} \\ 6,4\,0\,{}^{1}3 \\ -\ \ \ 2\,5\,7 \\ \hline 6,1\,4\,6 \end{array}
\qquad
\begin{array}{r} 1\,1 \\ 2\,5\,7 \\ +\,6,1\,4\,6 \\ \hline 6,4\,0\,3 \end{array}
$$

8.
$$
\begin{array}{r} {}^{6} \\ 8,7\,{}^{1}6\,5 \\ -\,3,0\,8\,5 \\ \hline 5,6\,8\,0 \end{array}
\qquad
\begin{array}{r} 1 \\ 3,0\,8\,5 \\ +\,5,6\,8\,0 \\ \hline 8,7\,6\,5 \end{array}
$$

9.
$$
\begin{array}{r} 4,9\,8\,7 \\ -\,3,7\,3\,2 \\ \hline 1,2\,5\,5 \end{array}
\qquad
\begin{array}{r} 3,7\,3\,2 \\ +\,1,2\,5\,5 \\ \hline 4,9\,8\,7 \end{array}
$$

10. $\$1,579 - \$890 = \$689$

11. $3,451 - 2,999 = 452$ miles

12. $1,976 - 1,917 = 59$ years

Lesson Practice 26C

1.
```
    4  1
 1,6 5 10      943
 -  9 4 3    + 707
    7 0 7    1,650
```

2.
```
   7 19       111
 8, 2 0 10    2,8 17
 -2,8 1 7    +5,383
   5,3 8 3    8,200
```

3.
```
   4 1 1       11
 5 2 121      4,740
 -4,7 4 0     + 481
     4 8 1    5,221
```

4.
```
    6 15       11
 8, 7 6 15     678
  -  6 7 8    +8,087
   8,0 8 7    8,765
```

5.
```
    8 9        11
 2, 9 0 16    1,088
 -1,0 8 8     +1,8 18
   1,8 1 8    2,906
```

6.
```
   5 13        11
 6 4 129      3,587
 -3,5 8 7     +2,842
   2,8 4 2    6,429
```

7.
```
    3 9        11
 3, 4 0 15     159
  -  1 5 9    +3,246
   3,2 4 6    3,405
```

8.
```
   6 9         11
 7, 0 01      5,99 1
 -5,9 9 1     +1,0 10
   1,0 1 0    7,001
```

9.
```
   8 12 1 1     111
 9, 3 2 14    8,358
 -8,3 5 8     + 966
     9 6 6    9,324
```

10. $2{,}004 - 1{,}492 = 512$ years
11. $\$3{,}600 - \$150 = \$3{,}450$
12. $2{,}562 - 1{,}600 = 962$ bushels

Systematic Review 26D

1.
```
   2 9         11
 4, 3 0 15     289
  -  2 8 9   +4,0 16
   4,0 1 6    4,305
```

2.
```
    4  3        1 1
 5, 8 4 10    3,9 14
 -3,9 1 4    +1,926
   1,9 2 6    5,840
```

3.
```
   6 9          11
 7, 0 113     1,523
 -1,5 2 3     +5,490
   5,4 9 0    7,013
```

4.
```
   1 9         11
 2 0 10        34
  -  3 4      +166
     1 6 6    200
```

5.
```
    6          1
 7 7 12       4 16
 -4 1 6      +356
   3 5 6      772
```

6.
```
   29          15
  -15         +14
   14          29
```

7. ⊺⊦⊣ |||
8. ⊺⊦⊣ ⊺⊦⊣ ⊺⊦⊣ ⊺⊦⊣ ⊺⊦⊣ ⊺⊦⊣ |
9. ⊺⊦⊣ ⊺⊦⊣ ||||
10. $6:50$
11. $3:14$
12. Tuesday
13. twelfth
14. $12 + 12 + 12 + 12 + 12 = 60"$
15. $451 + 385 = 836$ traveled
    $1{,}000 - 836 = 164$ to go

Systematic Review 26E

1.
```
   7 9          11
 1, 8 0 10     176
  -  1 7 6    +1,624
   1,6 2 4    1,800
```

2. $\begin{array}{r} 8\ ^14 \\ 6,\cancel{9}\ \cancel{5}\ ^10 \\ -4,8\ 6\ 7 \\ \hline 2,0\ 8\ 3 \end{array}$ $\begin{array}{r} 11 \\ 4,867 \\ +2,083 \\ \hline 6,950 \end{array}$

3. $\begin{array}{r} 8 \\ 2,0\cancel{9}\ ^13 \\ -2,0\ 7\ 5 \\ \hline 1\ 8 \end{array}$ $\begin{array}{r} 1 \\ 2,075 \\ +\quad 18 \\ \hline 2,093 \end{array}$

4. $\begin{array}{r} 6\ ^13 \\ \cancel{7}\ \cancel{4}\ ^11 \\ -\quad 5\ 3 \\ \hline 6\ 8\ 8 \end{array}$ $\begin{array}{r} 11 \\ 53 \\ +688 \\ \hline 741 \end{array}$

5. $\begin{array}{r} 4\ 9 \\ \cancel{5}\ \cancel{0}\ ^10 \\ -2\ 1\ 1 \\ \hline 2\ 8\ 9 \end{array}$ $\begin{array}{r} 11 \\ 211 \\ +289 \\ \hline 500 \end{array}$

6. $\begin{array}{r} 3 \\ \cancel{4}\ ^18 \\ -1\ 9 \\ \hline 2\ 9 \end{array}$ $\begin{array}{r} 1 \\ 19 \\ +29 \\ \hline 48 \end{array}$

7. 𝍩 (5 tally marks)

8. 𝍩 𝍩 𝍩 𝍩 IIII (tally marks totaling 24)

9. 𝍩 𝍩 II (tally marks totaling 12)

10. 11:39

11. 9:26

12. October

13. seventh

14. $673 + 415 = 1,088$
$1,088 - 223 = 865$ acorns

15. $10 + 10 + 10 + 10 = 40"$

Systematic Review 26F

1. $\begin{array}{r} 8\ 9 \\ 8,\cancel{9}\ \cancel{0}\ ^11 \\ -\quad 7\ 9\ 2 \\ \hline 8,1\ 0\ 9 \end{array}$ $\begin{array}{r} 11 \\ 792 \\ +8,109 \\ \hline 8,901 \end{array}$

2. $\begin{array}{r} 8\ 9\ ^10 \\ \cancel{9},\cancel{0}\ \cancel{1}\ ^12 \\ -8,9\ 9\ 9 \\ \hline 1\ 3 \end{array}$ $\begin{array}{r} 111 \\ 8,999 \\ +\quad 13 \\ \hline 9,012 \end{array}$

3. $\begin{array}{r} 1\ ^12 \\ 1,\cancel{2}\ \cancel{3}\ ^14 \\ -1,0\ 4\ 5 \\ \hline 1\ 8\ 9 \end{array}$ $\begin{array}{r} 11 \\ 1,045 \\ +\quad 189 \\ \hline 1,234 \end{array}$

4. $\begin{array}{r} 6 \\ 6\cancel{7}\ ^10 \\ -\quad 6\ 9 \\ \hline 6\ 0\ 1 \end{array}$ $\begin{array}{r} 1 \\ 69 \\ +601 \\ \hline 670 \end{array}$

5. $\begin{array}{r} 6\ ^15 \\ \cancel{7}\ \cancel{6}\ ^15 \\ -5\ 7\ 8 \\ \hline 1\ 8\ 7 \end{array}$ $\begin{array}{r} 11 \\ 578 \\ +187 \\ \hline 765 \end{array}$

6. $\begin{array}{r} 2 \\ \cancel{3}\ ^12 \\ -2\ 5 \\ \hline 7 \end{array}$ $\begin{array}{r} 1 \\ 25 \\ +\ 7 \\ \hline 32 \end{array}$

7. 𝍩 𝍩 I (tally marks totaling 11)

8. 𝍩 𝍩 𝍩 𝍩 𝍩 𝍩 III (tally marks totaling 33)

9. 𝍩 𝍩 𝍩 𝍩 𝍩 I (tally marks totaling 26)

10. 11:31

11. 7:41

12. Thursday

13. sixth

14. $2 + 4 + 5 + 6 + 8 = 25$ calls

15. $6 + 3 + 6 + 3 = 18"$

Lesson Practice 27A

1. done

2. $\begin{array}{r} 8\ 9 \\ \$\cancel{9}.\cancel{0}\ \cancel{0}\ ^10 \\ -2.6\ 7 \\ \hline \$6.3\ 3 \end{array}$

3. $\begin{array}{r} \$.46 \\ -.10 \\ \hline \$.36 \end{array}$

4. $\begin{array}{r} 5 \\ \$\cancel{6}.\ ^14 \\ -1.2\ 1 \\ \hline \$4.93 \end{array}$

5. $\begin{array}{r} 3 \\ \$2.\cancel{4}\ ^15 \\ -1.3\ 8 \\ \hline \$1.07 \end{array}$

6.
```
      7
  $.8 ¹0
  - .2 4
  $.5 6
```

7.
```
      4
  $5.¹19
  - 3.72
  $1.47
```

8.
```
    5 9
  $26.0¹0
  -11.9 2
  $14.0 8
```

9.
```
     3
  $34.4¹7
  -22.0 9
  $12.3 8
```

10. $7.15 − $2.98 = $4.17

11. $9.46 − $4.91 = $4.55

12. $45.50 − $34.99 = $10.51

Lesson Practice 27B

1.
```
     5
  $2.6¹5
  - .3 8
  $2.2 7
```

2.
```
   6 ¹1
  $7.2¹5
  - 1.8 9
  $5.3 6
```

3.
```
  $.10
  -.08
  $.02
```

4.
```
     9
  $1.0¹0
  - .7 7
  $ .2 3
```

5.
```
     5
  $6.¹19
  - 3.58
  $2.6 1
```

6.
```
    8
  $.9¹3
  -.4 5
  $.4 8
```

7.
```
     3
  $2.4¹7
  -1.0 9
  $1.3 8
```

8.
```
     4
  $35.¹60
  -21. 90
  $13. 70
```

9.
```
   6  7
  $74.8¹2
  - 3 6.2 5
  $ 3 8.5 7
```

10. $31.16 − $25.69 = $5.47

11. $39.67 − $20.75 = $18.92

12. $13.05 − $.36 = $12.69

Lesson Practice 27C

1.
```
   4 ¹0
  $5.1¹3
  - .4 8
  $4.6 5
```

2.
```
     0
  $8.1¹5
  -7.0 9
  $1.0 6
```

3.
```
    6
  $.7¹3
  -.1 7
  $.5 6
```

4.
```
   2 9
  $3.0¹0
  - .8 1
  $2.1 9
```

5.
```
    2
  $9.3¹2
  -6.1 4
  $3.1 8
```

6.
```
  $.86
  -.55
  $.31
```

7.
```
    6
  $4.7¹5
  -2.0 6
  $2.6 9
```

8.
$$\begin{array}{r} 1\,9 \\ \$\cancel{2}\,\cancel{0}.{}^{1}15 \\ -\ 13.\ 2\,1 \\ \hline \$\ \ 6.\ 9\,4 \end{array}$$

9.
$$\begin{array}{r} 7\ {}^{1}0 \\ \$9\,\cancel{8}.\,\cancel{1}\,{}^{1}7 \\ -25.\ 1\,8 \\ \hline \$72.\ 9\,9 \end{array}$$

10. $10.00 - 1.25 = 8.75$

11. $53.95 - 45.00 = 8.95$

12. $.61 - .45 = .16$

Systematic Review 27D

1.
$$\begin{array}{r} 8 \\ \$8.\cancel{9}\,{}^{1}1 \\ -\ .7\,6 \\ \hline \$8.\ 1\,5 \end{array}$$

2.
$$\begin{array}{r} \$6.82 \\ -1.6\,1 \\ \hline \$5.2\,1 \end{array}$$

3.
$$\begin{array}{r} 6\ {}^{1}6 \\ \cancel{7},\cancel{7}\,{}^{1}12 \\ -5,8\,7\,2 \\ \hline 1,\ 8\,4\,0 \end{array}$$

4.
$$\begin{array}{r} 1\,1 \\ 184 \\ +\ 68 \\ \hline 252 \end{array}$$

5.
$$\begin{array}{r} 1\,1 \\ 255 \\ +177 \\ \hline 432 \end{array}$$

6.
$$\begin{array}{r} 1 \\ 3,011 \\ +1,895 \\ \hline 4,906 \end{array}$$

7. first — Wednesday
8. second — Friday
9. third — Saturday
10. fourth — Sunday
11. fifth — Thursday
12. sixth — Monday
13. seventh — Tuesday

14. 2, 4, 6, 8, 10, 12, 14, 16, 18, 20
15. 17 birds
16. $5 + 5 + 5 + 5 = 20$
     $20.00 - 6.95 = 13.05$

Systematic Review 27E

1.
$$\begin{array}{r} 3 \\ \$4.\cancel{4}\,{}^{1}2 \\ -\ .3\,9 \\ \hline \$4.0\,3 \end{array}$$

2.
$$\begin{array}{r} 1 \\ \$1.\cancel{2}\,{}^{1}0 \\ -1.1\,8 \\ \hline \$\ .0\,2 \end{array}$$

3.
$$\begin{array}{r} 3 \\ \cancel{4},{}^{1}503 \\ -2,9\,0\,1 \\ \hline 1,\ 6\,02 \end{array}$$

4.
$$\begin{array}{r} 1 \\ 836 \\ +\ 17 \\ \hline 853 \end{array}$$

5.
$$\begin{array}{r} 1 \\ \$\ 3.45 \\ +\ 7.18 \\ \hline \$10.63 \end{array}$$

6.
$$\begin{array}{r} 1\ \ 1 \\ 1,819 \\ +4,428 \\ \hline 6,247 \end{array}$$

7. first — June
8. second — April
9. third — January
10. fourth — March
11. fifth — February
12. sixth — May

13. 5, 10, 15, 20, 25, 30, 35, 40, 45, 50
14. 30 days
15. Drew: $5.00 + 6.50 = 11.50$
     $11.50 > 10.20$; Drew earned more.

Systematic Review 27F

1. $$\begin{array}{r} 0 \\ \$3.\cancel{1}\,^14 \\ -1.0\;9 \\ \hline \$2.0\;5 \end{array}$$

2. $$\begin{array}{r} 5 \\ \$2.\cancel{6}\,^12 \\ -1.3\;8 \\ \hline \$1.2\;4 \end{array}$$

3. $$\begin{array}{r} 8 \\ \$1\cancel{9}.\,^107 \\ -15.\;21 \\ \hline \$\;\;3.\;86 \end{array}$$

4. $$\begin{array}{r} 1 \\ 977 \\ +\;\;13 \\ \hline 990 \end{array}$$

5. $$\begin{array}{r} 1 \\ \$\;6.54 \\ +\;9.54 \\ \hline \$16.08 \end{array}$$

6. $$\begin{array}{r} 2\;1\;1 \\ 1,764 \\ 1,913 \\ +2,384 \\ \hline 6,061 \end{array}$$

7. seventh — August
8. eighth — October
9. ninth — December
10. tenth — July
11. eleventh — November
12. twelfth — September
13. 10, 20, 30, 40, 50, 60, 70, 80, 90, 100
14. ⅲ
15. $35+25+19 = 79'$
    $79-10 = 69'$

Lesson Practice 28A

1. done

2. $$\begin{array}{r} 84,528 \\ -64,025 \\ \hline 20,503 \end{array} \qquad \begin{array}{r} 64,025 \\ +20,503 \\ \hline 84,528 \end{array}$$

3. $$\begin{array}{r} 4\quad 8 \\ 3\cancel{5},\!^1\cancel{19}\,^14 \\ -31,486 \\ \hline 3,708 \end{array} \qquad \begin{array}{r} 1\;\;1 \\ 31,486 \\ +\;3,708 \\ \hline 35,194 \end{array}$$

4. $$\begin{array}{r} 1\;\,^12 \\ 4\cancel{2},\!\cancel{3}\,^155 \\ -21,472 \\ \hline 20,883 \end{array} \qquad \begin{array}{r} 1\;1 \\ 21,472 \\ +20,883 \\ \hline 42,355 \end{array}$$

5. $$\begin{array}{r} 5\,^11 \\ 71,\!\cancel{6}\cancel{2}\,^11 \\ -41,573 \\ \hline 30,048 \end{array} \qquad \begin{array}{r} 1\;1 \\ 41,573 \\ +30,048 \\ \hline 71,621 \end{array}$$

6. $$\begin{array}{r} 3\;9 \\ 56,\!\cancel{4}\cancel{0}\,^18 \\ -24,379 \\ \hline 32,029 \end{array} \qquad \begin{array}{r} 1\;1 \\ 24,379 \\ +32,029 \\ \hline 56,408 \end{array}$$

7. $45,900 - 22,175 = 23,725$ people
8. $31,231 - 19,452 = 11,779$ tadpoles
9. $25,000 - 19,000 = 6,000$ miles
10. $\$45,575 - \$38,196 = \$7,379$

Lesson Practice 28B

1. $$\begin{array}{r} 5\quad 4\,^10 \\ \cancel{6}\,^18,\!\cancel{5}\,^1\cancel{1}\,^11 \\ -19,333 \\ \hline 49,178 \end{array} \qquad \begin{array}{r} 1\;11 \\ 19,333 \\ +49,178 \\ \hline 68,511 \end{array}$$

2. $$\begin{array}{r} 7 \\ 25,3\cancel{8}\,^12 \\ -11,255 \\ \hline 14,127 \end{array} \qquad \begin{array}{r} 1 \\ 11,255 \\ +14,127 \\ \hline 25,382 \end{array}$$

3. $$\begin{array}{r} 5 \\ 47,4\cancel{6}\,^10 \\ -33,419 \\ \hline 14,041 \end{array} \qquad \begin{array}{r} 1 \\ 33,419 \\ +14,041 \\ \hline 47,460 \end{array}$$

4. $$\begin{array}{r} 6\;9\;9\;9 \\ \cancel{7}\cancel{0},\!\cancel{0}\cancel{0}\,^10 \\ -19,999 \\ \hline 50,001 \end{array} \qquad \begin{array}{r} 1\;1\;1\;1 \\ 19,999 \\ +50,001 \\ \hline 70,000 \end{array}$$

5. $$\begin{array}{r} 3 \\ 65,2\cancel{4}\,^12 \\ -21,135 \\ \hline 44,107 \end{array} \qquad \begin{array}{r} 1 \\ 21,135 \\ +44,107 \\ \hline 65,242 \end{array}$$

6.
$$
\begin{array}{r}
5 \\
54,9\overset{1}{\cancel{6}}7 \\
-42,718 \\
\hline
12,249
\end{array}
\qquad
\begin{array}{r}
1 \\
42,718 \\
+12,249 \\
\hline
54,967
\end{array}
$$

7. $10,752 - 6,834 = 3,918$ people
8. $55,780 - 29,592 = 26,188$ plants
9. $63,360 - 31,680 = 31,680$ inches
10. $\$22,980 - \$16,899 = \$6,081$

Lesson Practice 28C

1.
$$
\begin{array}{r}
4\,{}^{1}1 \\
13,\overset{1}{\cancel{5}}\,\overset{}{\cancel{2}}\,2 \\
-12,048 \\
\hline
1,474
\end{array}
\qquad
\begin{array}{r}
11 \\
12,048 \\
+1,474 \\
\hline
13,522
\end{array}
$$

2.
$$
\begin{array}{r}
8\,{}^{1} \\
7\overset{1}{\cancel{9}},147 \\
-24,312 \\
\hline
54,835
\end{array}
\qquad
\begin{array}{r}
1 \\
24,312 \\
+54,835 \\
\hline
79,147
\end{array}
$$

3.
$$
\begin{array}{r}
5\,{}^{1}3 \\
53,\overset{}{\cancel{6}}\,\overset{1}{\cancel{4}}\,1 \\
-20,465 \\
\hline
33,176
\end{array}
\qquad
\begin{array}{r}
11 \\
20,465 \\
+33,176 \\
\hline
53,641
\end{array}
$$

4.
$$
\begin{array}{r}
7\,{}^{1}3\;\;6 \\
8\,\overset{}{\cancel{4}},6\,\overset{1}{\cancel{7}}\,\overset{1}{\cancel{0}} \\
-35,901 \\
\hline
48,769
\end{array}
\qquad
\begin{array}{r}
11\;1 \\
35,901 \\
+48,769 \\
\hline
84,670
\end{array}
$$

5.
$$
\begin{array}{r}
{}^{1}2\,{}^{1}4 \\
\overset{}{\cancel{1}}\,3,\overset{}{\cancel{5}}\,{}^{1}06 \\
-9,951 \\
\hline
3,555
\end{array}
\qquad
\begin{array}{r}
111 \\
9,951 \\
+3,555 \\
\hline
13,506
\end{array}
$$

6.
$$
\begin{array}{r}
0\;9 \\
4\,\overset{}{\cancel{1}},\overset{}{\cancel{0}}\,{}^{1}03 \\
-10,523 \\
\hline
30,480
\end{array}
\qquad
\begin{array}{r}
11 \\
10,523 \\
+30,480 \\
\hline
41,003
\end{array}
$$

7. $23,295 - 19,761 = 3,534$ miles
8. $53,600 - 48,096 = 5,504$ blocks
9. $60,000 - 22,000 = 38,000$ fish
10. $\$95,899 - \$26,900 = \$68,999$

Systematic Review 28D

1.
$$
\begin{array}{r}
5\,9\,{}^{1}1 \\
9\overset{1}{\cancel{6}},0\,\overset{}{\cancel{2}}\,{}^{1}1 \\
-45,635 \\
\hline
50,386
\end{array}
\qquad
\begin{array}{r}
111 \\
45,635 \\
+50,386 \\
\hline
96,021
\end{array}
$$

2.
$$
\begin{array}{r}
7\,9\,9\,9 \\
8\,\overset{}{\cancel{0}},\overset{}{\cancel{0}}\,\overset{}{\cancel{0}}\,{}^{1}0 \\
-79,998 \\
\hline
2
\end{array}
\qquad
\begin{array}{r}
1\,1\,1\,1 \\
79,998 \\
+\quad\;\;\;2 \\
\hline
80,000
\end{array}
$$

3.
$$
\begin{array}{r}
\$50.75 \\
-10.25 \\
\hline
\$40.50
\end{array}
$$

4.
$$
\begin{array}{r}
8\,{}^{1}0 \\
\$3\overset{}{\cancel{9}}.\,\overset{}{\cancel{1}}\,{}^{1}3 \\
-22.46 \\
\hline
\$16.67
\end{array}
$$

5.
$$
\begin{array}{r}
8\,{}^{1}0 \\
\$1\overset{}{\cancel{9}}.\,\overset{}{\cancel{1}}\,{}^{1}5 \\
-7.99 \\
\hline
\$11.16
\end{array}
$$

6. $3 < 4$
7. $7 > 6$
8. $9 = 9$

9.
$$
\begin{array}{r}
21 \\
125 \\
285 \\
44 \\
+161 \\
\hline
615
\end{array}
$$

10.
$$
\begin{array}{r}
11 \\
315 \\
200 \\
394 \\
+155 \\
\hline
1,064
\end{array}
$$

11.
$$
\begin{array}{r}
12 \\
449 \\
131 \\
503 \\
+\;\;67 \\
\hline
1,150
\end{array}
$$

12. Sunday
13. third
14. $\$123.45 + \$210.13 + \$75.21 + \$103.82 = \$512.61$
15. $100 - 99 = 1$ sheep

Systematic Review 28E

1.
$$\begin{array}{r} 0\ ^13\ \ \ \\ 1\!\!1,\!\cancel{4}\ ^135 \\ -10,\!682 \\ \hline 753 \end{array} \qquad \begin{array}{r} 1\ 1\ \\ 10,\!682 \\ +\ \ \ 753 \\ \hline 11,\!435 \end{array}$$

2.
$$\begin{array}{r} 3\ \ 6\ ^12\ \ \\ \cancel{4}\ ^10,\!\cancel{7}\ \cancel{3}\ ^14 \\ -27,\!156 \\ \hline 13,\!578 \end{array} \qquad \begin{array}{r} 1\ 11\ \\ 27,\!156 \\ +13,\!578 \\ \hline 40,\!734 \end{array}$$

3.
$$\begin{array}{r} 1\ ^10\ ^10\ \ \\ \$\cancel{2}\ \cancel{1}.\ \cancel{1}\ ^11 \\ -\ 19.\ 89 \\ \hline \$\ 1.\ 22 \end{array}$$

4.
$$\begin{array}{r} 7\ \ \\ \$7\cancel{8}.\ ^104 \\ -35.\ 40 \\ \hline \$42.\ 64 \end{array}$$

5.
$$\begin{array}{r} 1\ 9\ \\ \$6\cancel{2}.\cancel{0}\ ^10 \\ -51.\ 19 \\ \hline \$10.\ 8\ 1 \end{array}$$

6. $14 = 14$

7. $2 < 3$

8. $63 > 36$

9.
$$\begin{array}{r} 1\ \ \\ 704 \\ 300 \\ 26 \\ +\ \ 9 \\ \hline 1,\!039 \end{array}$$

10.
$$\begin{array}{r} 12\ \ \\ 241 \\ 119 \\ 562 \\ +508 \\ \hline 1,\!430 \end{array}$$

11.
$$\begin{array}{r} 11\ \ \\ 325 \\ 763 \\ 125 \\ +\ 46 \\ \hline 1,\!259 \end{array}$$

12. August

13. sixth

14. $10,\!456 - 8,\!912 = 1,\!544$ spiderlings

15. $25 + 89 + 5 = 119$ animals

Systematic Review 28F

1.
$$\begin{array}{r} 2\ \ \ \ \ \\ 79,\!\cancel{3}\ ^104 \\ -61,\!082 \\ \hline 18,\!222 \end{array} \qquad \begin{array}{r} 1\ \ \ \ \\ 61,\!082 \\ +18,\!222 \\ \hline 79,\!304 \end{array}$$

2.
$$\begin{array}{r} 4\ ^1499\ \\ \cancel{5}\ \cancel{5},\!\cancel{0}\ \cancel{0}\ ^10 \\ -48,\!123 \\ \hline 6,\!877 \end{array} \qquad \begin{array}{r} 1111\ \\ 48,\!123 \\ +\ 6,\!877 \\ \hline 55,\!000 \end{array}$$

3.
$$\begin{array}{r} \$\ 17.99 \\ -\ \ 5.15 \\ \hline \$\ 12.84 \end{array}$$

4.
$$\begin{array}{r} 3\ ^12\ ^15 \\ \$4\cancel{3}.\cancel{6}\ ^10 \\ -\ 13.\ 79 \\ \hline \$\ 29.\ 8\ 1 \end{array}$$

5.
$$\begin{array}{r} 6\ \ \\ \$28.\cancel{7}\ ^14 \\ -\ 13.\ 2\ 5 \\ \hline \$\ 15.\ 4\ 9 \end{array}$$

6. $42 < 78$

7. $16 > 11$

8. $3 = 3$

9.
$$\begin{array}{r} 11\ \ \\ 910 \\ 262 \\ 366 \\ +148 \\ \hline 1,\!686 \end{array}$$

10.
$$\begin{array}{r} 21\ \ \\ 355 \\ 176 \\ 755 \\ +\ 23 \\ \hline 1,\!309 \end{array}$$

11.
$$\begin{array}{r} 11\ \ \\ 431 \\ 529 \\ 611 \\ +576 \\ \hline 2,\!147 \end{array}$$

12. Wednesday

13. fifth

14. $\$34.00 + \$25.00 = \$59.00$

$\$59.00 - \$29.98 = \$29.02$

15. $25 + 17 + 30 + 8 = 80$ cars

Lesson Practice 29A

1. done
2. 5 (See example 1 on student page for filled in numbers on gauge.)
3. 10
4. 50
5. 10
6. 70
7. done

8.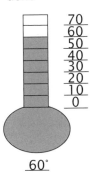

   60°

9.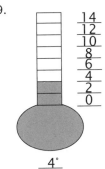

   4°

7.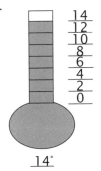

   14°

8.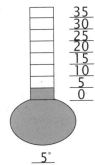

   5°

9.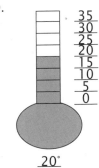

   20°

Lesson Practice 29B

1. done
2. 100°
3. 600°
4. 30
5. 90
6. 60

Lesson Practice 29C

1. 20
2. 0
3. 15
4. 20
5. 80
6. 40

7.

$\underline{30°}$

8.

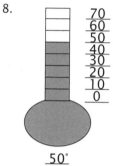

$\underline{50°}$

9.

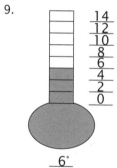

$\underline{6°}$

3.

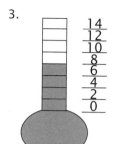

$\underline{8°}$

4. 100

5. 10

6.
$$\begin{array}{r} 2\,{}^13\,{}^10\,{}^11 \\ \cancel{3}\,\cancel{4},\,\cancel{1}\,\cancel{2}\,{}^17 \\ -2\,5,\,1\,7\,9 \\ \hline 8,\,9\,4\,8 \end{array}$$

7.
$$\begin{array}{r} {}^7 \\ 5,6\,\cancel{8}\,{}^10 \\ -4,4\,5\,2 \\ \hline 1,2\,2\,8 \end{array}$$

8.
$$\begin{array}{r} 1\,1 \\ 7,5\,6\,3 \\ +5,1\,4\,8 \\ \hline 12,7\,1\,1 \end{array}$$

9. 9:35

10. 1:16

11. 41 years

12. ||||| ||||| ||||| ||||| |||||

 ||||| ||||| ||||| |

Systematic Review 29D

1. 5

2. 300°

Systematic Review 29E

1. 20

2. 500°

3.

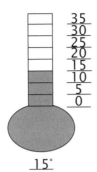

$\underline{15°}$

4. 50

5. 0

6.
$$\begin{array}{r} 6\ {}^14 \\ 1\cancel{7},\cancel{5}\ {}^139 \\ -15,6\ 5\ 9 \\ \hline 1,8\ 8\ 0 \end{array}$$

7.
$$\begin{array}{r} 6\ 9 \\ \cancel{7},\cancel{0}\ {}^158 \\ -3,1\ 7\ 2 \\ \hline 3,8\ 8\ 6 \end{array}$$

8.
$$\begin{array}{r} 1\ 1 \\ 4,264 \\ +2,791 \\ \hline 7,055 \end{array}$$

9. left

10. $1976 - 200 = 1776$; seventh

Systematic Review 29F

1. ~~10~~
2. ~~200°~~

1. 10

2. 200°

3.

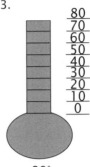

$\underline{80°}$

4. 10

5. 60

6.
$$\begin{array}{r} 2\ {}^109 \\ 9\cancel{3},\cancel{1}\cancel{0}\ {}^15 \\ -60,1\ 2\ 7 \\ \hline 3\ 2,9\ 7\ 8 \end{array}$$

7.
$$\begin{array}{r} 7\ 9\ 9 \\ \cancel{8},\cancel{0}\cancel{0}\ {}^10 \\ -2,1\ 1\ 1 \\ \hline 5,8\ 8\ 9 \end{array}$$

8.
$$\begin{array}{r} 1\ 1\ 1 \\ 6,456 \\ 2,434 \\ +1,849 \\ \hline 10,739 \end{array}$$

9. 11 : 43

10. 8 : 26

11. left

12. step 1: $1976 - 200 = 1776$

step 2: Subtract 1776 from current year.
Answers will vary according to year.
There is more than one correct way
to solve this.

Lesson Practice 30A

1. February
2. December
3. 2"
4. January
5.

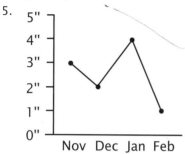

6. Carol
7. 7
8. 10
9. Adam and Ellen
10.

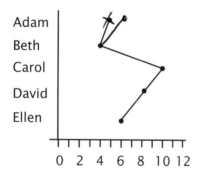

Lesson Practice 30B

1. Wednesday
2. Tuesday
3. 200 donuts
4. 300 donuts

5.

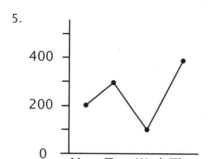

6. blue
7. 6 cars
8. 5 cars
9. 2 cars
10.

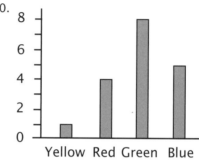

Lesson Practice 30C

1. 100 miles
2. 50 miles
3. $100 + 200 + 150 + 250 + 50 = 750$
4.

250
200
150
100
50
0

Mon Tue Wed Thu Fri

5. 16 books
6. 12 books
7. $8+12+10+16 = 46$ books
8.

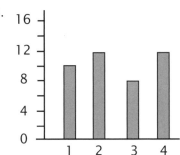

10. right
11. 26
12. March

Systematic Review 30E

1. May
2. July
3. 5"
4.

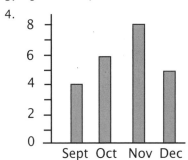

5. 80
6. 40
7. $135 < 531$
8. $15 = 15$
9. $8 < 9$
10. $45+45+45+45 = 180'$
11. $1972-1945 = 27$ years
12. $65+348+179 = 592$ miles

Systematic Review 30D

1. Thursday and Friday
2. Wednesday
3. $3+3+2+4+4 = 16$ miles
4.

4 ●
       ●   ●
2   ●
          ●
0
  Mon Tue Wed Thu Fri

5. 15
6. 100°

7.
$$\begin{array}{r} 7\;^18 \\ \$7\,\cancel{8}.\cancel{9}\,^10 \\ -\,21.\;9\;5 \\ \hline \$5\,6.\;9\;5 \end{array}$$

8.
$$\begin{array}{r} 1\;1 \\ \$\,13.5\,4 \\ +\;\;\;8.6\,3 \\ \hline \$\,22.1\,7 \end{array}$$

9.
$$\begin{array}{r} 4\,^12\,^13\,^10 \\ \cancel{5}\,\cancel{3},\cancel{4}\,\cancel{1}\,^11 \\ -\,16,\,4\;2\;2 \\ \hline 3\,6,\,9\;8\;9 \end{array}$$

Systematic Review 30F

1. Tigers
2. Red Sox
3. 9 runs
4. Orioles and Blue Jays

5.

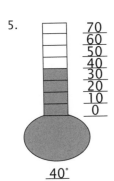

40°

6.

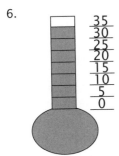

35°

7.

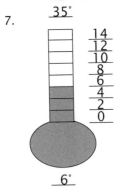

6°

8.
$$
\begin{array}{r}
4\,{}^{1}3\\
76,5\,4\,{}^{1}1\\
-32,1\,5\,3\\
\hline
44,3\,8\,8
\end{array}
$$

9.
$$
\begin{array}{r}
5\,9\,9\\
6,0\,0\,{}^{1}0\\
-3,2\,6\,5\\
\hline
2,7\,3\,5
\end{array}
$$

10.
$$
\begin{array}{r}
1\,1\,1\\
3,128\\
5,592\\
+4,517\\
\hline
13,237
\end{array}
$$

Test Solutions

Test 1

1. 264; "two hundred sixty-four"
2. 39; "thirty-nine"
3. 4 hundreds, 2 tens, and 1 unit; "four hundred twenty-one"
4. 2 hundreds and 7 units; "two hundred seven"
5. $0 + 7 = 7$
6. $8 + 3 = 11$
7. $9 + 7 = 16$
8. $6 + 5 = 11$
9. $4 + 4 = 8$
10. $3 + 6 = 9$
11. $8 + 2 = 10$
12. $4 + 7 = 11$
13. $5 + 5 = 10$
14. $3 + 5 = 8$
15. $6 + 4 = 10$
16. $5 + 7 = 12$
17. $9 + 1 = 10$
18. $9 + 6 = 15$

Test 2

1. 16, 20, 34
2. 19, 49, 99
3. 108, 83, 82
4. 111, 101, 11
5. 22, 23, 24, 25, 26
6. 1 hundred, and 4 tens "one hundred forty"
7. 2 tens and 2 units "twenty-two"
8. $4 + 0 = 4$
9. $2 + 7 = 9$
10. $5 + 9 = 14$
11. $3 + 6 = 9$
12. $2 + 3 = 5$
13. $7 + 7 = 14$
14. $9 + 1 = 10$
15. 14 is less than 24, so Chad
16. $8 + 9 = 17$ arrowheads

Test 3

1. $11 > 7$
2. $9 = 9$
3. $19 < 91$
4. $111 > 101$
5. $13 = 13$
6. $9 < 10$
7. 29, 28, 27, 26, 25
8. $2 + 8 = 10$
9. $6 + 9 = 15$
10. $3 + 4 = 7$
11. $0 + 2 = 2$
12. $7 + 8 = 15$
13. $4 + 7 = 11$
14. $9 + 4 = 13$
15. $12 < 14$
16. $5 + 0 = 5$ dollars

Test 4

1. 90
2. 30
3. 50
4.
$$\begin{array}{r}(50)\\ +(20)\\ \hline (70)\end{array}$$
5.
$$\begin{array}{r}(40)\\ +(30)\\ \hline (70)\end{array}$$
6.
$$\begin{array}{r}(30)\\ +(30)\\ \hline (60)\end{array}$$
7.
$$\begin{array}{r}(80)\\ +(10)\\ \hline (90)\end{array}$$
8.

4	4	8
3	2	5
7	6	13

9.

1	6	7
5	3	8
6	9	15

10. $10 > 7$
11. $399 > 329$

12. $2+2 = 4$ muffins

13. $(30)+(30) = (60)$ cans

14. $201 > 21$

Test 5

1. $600+20+8$

2. $40+9$

3.   16   10+6
  +32 +30+2
   48   40+8

4.   380   300+80+0
 +516 +500+10+6
  896   800+90+6

5.   54   50+4
 +22 +20+2
  76   70+6

6.   433   400+30+3
 +425 +400+20+5
  858   800+50+8

7.   34   (30)
 +61 +(60)
  95   (90)

8.   12   (10)
 +16 +(20)
  28   (30)

9. $10 = 10$

10. $49 < 94$

11. $7+\underline{7} = 14$

12. $5+\underline{3} = 8$

13. $8+\underline{8} = 16$

14. $45+12 = 57$ animals

15. $132+160 = 292$ coins

Test 6

1. 2, 4, 6, 8, 10, 12, 14, 16, 18, 20

2. 2, 4, 6, $\underline{8}$ cones

3.   55   50+5
 + 2 +00+2
  57   50+7

4.   106   100+00+6
 +221 +200+20+1
  327   300+20+7

5.   44   (40)
 +13 +(10)
  57   (50)

6.   38   (40)
 +21 +(20)
  59   (60)

7. $2+\underline{6} = 8$

8. $5+\underline{8} = 13$

9. $9+\underline{9} = 18$

10. $5+9 = 14$ trucks

11. 2, 4, $\underline{6}$ hands

12. $40+25 = 65$ jelly beans

Test 7

1.   1
  35
 +29
  64

2.   1
  16
 +67
  83

3.   1
  29
 +44
  73

4.   1
  19
 +46
  65

5.   1
  54
 + 7
  61

6.   75
 +22
  97

7. 2, 4, 6, 8, 10, 12, 14, 16, 18, 20

8. $15 > 13$
9. $21 > 12$
10. $801 < 810$
11. $4 + \underline{1} = 5$
12. $9 + \underline{3} = 12$
13. $4 + \underline{3} = 7$
14. $55 + 39 = 94$ quarts
15. $212 + 344 = 556$ animals

Unit Test I

1. 134
   "one hundred thirty-four"
2. 17, 71, 111
3. $11 > 10$
4. $14 = 14$
5. $342 > 324$
6. 20
7. 40
8. 80
9. $5 + \underline{2} = 7$
10. $8 + \underline{8} = 16$
11. $5 + \underline{4} = 9$
12. 2, 4, 6, 8, 10, 12, 14, 16, 18, 20
13. $\begin{array}{r} 16 \\ +21 \\ \hline 37 \end{array}$
14. $\begin{array}{r} 1 \\ 55 \\ +37 \\ \hline 92 \end{array}$
15. $\begin{array}{r} 1 \\ 15 \\ +25 \\ \hline 40 \end{array}$
16. $\begin{array}{r} 1 \\ 78 \\ +\ \ 8 \\ \hline 86 \end{array}$
17. $\begin{array}{r} 234 \\ +142 \\ \hline 376 \end{array}$
18. $\begin{array}{r} 361 \\ +205 \\ \hline 566 \end{array}$

19. 2, 4, 6, $\underline{8}$ letters
20. $(30) + (40) = (70)$
    $28 + 43 = 71$ cars
21. $415 + 221 = 636$ fish
22. $39 + 15 = 54$ dollars

Test 8

1. 10, 20, 30, 40, 50, 60, 70, 80, 90, 100
2. 2, 4, 6, 8, 10, 12, 14, 16, 18, 20
3. $\begin{array}{r} 1 \\ 45 \\ +27 \\ \hline 72 \end{array}$
4. $\begin{array}{r} 1 \\ 65 \\ +16 \\ \hline 81 \end{array}$
5. $\begin{array}{r} 1 \\ 37 \\ +34 \\ \hline 71 \end{array}$
6. $\begin{array}{r} 109 \\ +290 \\ \hline 399 \end{array}$
7. $\begin{array}{r} 477 \\ +122 \\ \hline 599 \end{array}$
8. $\begin{array}{r} 834 \\ +\ \ 45 \\ \hline 879 \end{array}$
9. $4 + \underline{4} = 8$
10. $5 + \underline{9} = 14$
11. $7 + \underline{3} = 10$
12. 10, 20, 30, 40, 50, 60, $\underline{70}$ toes
13. $4 + 1 = 5$ dimes
    10, 20, 30, 40, $\underline{50}$¢
14. $6 + 1 = 7$
    $7 + 3 = 10$ stories
15. $34 + 47 = 81$ children

Test 9

1. 5, 10, 15, 20, 25, 30, 35, 40, 45, 50
2. 10, 20, 30, 40, 50, 60, 70, 80, 90, 100

3.
```
   1
  26
 +55
  81
```

4.
```
   1
  68
 + 8
  76
```

5.
```
   1
  13
 +49
  62
```

6.
```
  116
 +431
  547
```

7.
```
  691
 +305
  996
```

8.
```
  500
 +216
  716
```

9. $7 + \underline{6} = 13$
10. $5 + \underline{4} = 9$
11. $5 + \underline{0} = 5$
12. penny
13. nickel
14. dime
15. 5, 10, 15, 20, 25, <u>30</u> rocks

Test 10

1. $2.34
   "two dollars and thirty-four cents"
2. 1 dollar, 2 dimes, and 8 pennies
   "one dollar and twenty-eight cents'
3. 2 dollars and 6 pennies
   "two dollars and six cents"
4. 3 dollars, 4 dimes, and one penny
   "three dollars and forty-one cents"
5. 4 dollars, 1 dime, and 5 pennies
   "four dollars and fifteen cents"

6. 5, 10, 15, 20, 25¢

7.
```
   1
  63
 +17
  80
```

8.
```
  234
 +145
  379
```

9.
```
   1
  48
 +26
  74
```

10. 50
11. 20
12. 30
13. $9.54
14. $3 + 2 = 5$
    $5 + 5 = 10$ pounds
15. 2 dimes = 20¢
    5 nickels = 25¢
    20¢ + 25¢ = 45¢ or $.45

Test 11

1. 100
2. 300
3. 500

4.
```
    1
  132
 +418
  550
```

5.
```
   11
  679
 +276
  955
```

6.
```
   1
  520
 +188
  708
```

7.
```
   1
  87
 + 28
 115
```

8.
```
  45
 +42
  87
```

9.
```
  1
  39
 +15
  54
```

10. $7 - 2 = 5$
11. $5 - 1 = 4$
12. $8 - 0 = 8$
13. $11 - 9 = 2$
14. $5 - 3 = 2$
15. $10 - 2 = 8$
16. $6 - 1 = 5$
17. $7 - 5 = 2$
18. 5, 10, 15, 20, 25, 30, 35, 40, 45, 50
19. $(500) + (100) = (600)$
    $476 + 125 = 601$ stamps
20. $10 - 6 = 4$ years

Test 12

1.
```
  1 1
 $2.46
 +1.79
 $4.25
```

2.
```
  1
 $3.91
 +3.25
 $7.16
```

3.
```
  1
 $1.57
 + .38
 $1.95
```

4.
```
  1
  401
 +579
  980
```

5.
```
  65
 +11
  76
```

6.
```
  1
  22
 + 8
  30
```

7. $18 - 9 = 9$
8. $8 - 4 = 4$
9. $13 - 8 = 5$
10. $16 - 9 = 7$
11. $10 - 5 = 5$
12. $12 - 8 = 4$
13. $14 - 7 = 7$
14. $13 - 9 = 4$
15. $1.43; "one dollar and forty-three cents"
16. $1.89 + $2.36 = $4.25
17. 4 dimes = $.40
    5 nickels = $.25
    $.40 + $.25 = $.65
18. $15 - 9 = 6$ people

Test 13

1. $6 + 4 + 7 = 17$
2. $3 + 2 + 7 + 8 = 20$
3.
```
  1
  11
  42
  39
 +36
  128
```

4.
```
  1
 $1.26
 +8.47
 $9.73
```

5.
```
  1
  641
 + 90
  731
```

6.
```
  1
  58
 + 2
  60
```

7. $10 - 7 = 3$
8. $9 - 3 = 6$
9. $10 - 6 = 4$
10. $9 - 4 = 5$
11. $10 - 5 = 5$
12. $9 - 7 = 2$

13. $10 - 2 = 8$
14. $10 - 3 = 7$
15. 2, 4, 6, 8, 10, 12, 14, 16, 18, 20
16. triangle
17. square
18. rectangle
19. $3 + 4 + 6 + 1 = 14$ hours
20. 6 nickels = 30¢
    3 dimes = 30¢
    $30¢ + 30¢ + 15¢ = 75¢$

Test 14

1. $3"$
2. $6"$
3.
$$\begin{array}{r} 1 \\ \$1.34 \\ +2.07 \\ \hline \$3.41 \end{array}$$
4.
$$\begin{array}{r} 11 \\ 866 \\ +\ 36 \\ \hline 902 \end{array}$$
5.
$$\begin{array}{r} 1 \\ 26 \\ +44 \\ \hline 70 \end{array}$$
6. $1 + 1 + 3 + 9 + 7 = 21$
7. $13 + 4 + 11 + 6 = 34$
8. $9 > 8$
9. $13 - 7 = 6$
10. $15 - 9 = 6$
11. $7 - 3 = 4$
12. $14 - 6 = 8$
13. $7 - 4 = 3$
14. $12 - 7 = 5$
15. 3 sides
16. 4 sides
17. $11 - 4 = 7$
    $7 - 2 = 5$ left
18. $12 + 12 + 12 + 12 = 48"$

Test 15

1. triangle
   $2 + 4 + 5 = 11"$
2. square
   $14 + 14 + 14 + 14 = 56"$
3.
$$\begin{array}{r} 1 \\ \$2.22 \\ +3.68 \\ \hline \$5.90 \end{array}$$
4.
$$\begin{array}{r} 1 \\ 771 \\ +135 \\ \hline 906 \end{array}$$
5.
$$\begin{array}{r} 1 \\ 46 \\ +55 \\ \hline 101 \end{array}$$
6. $13 - 9 = 4$
7. $11 - 4 = 7$
8. $13 - 4 = 9$
9. $12 - 5 = 7$
10. $9 - 6 = 3$
11. $11 - 3 = 8$
12. $5 + 6 + 5 + 6 = 22'$
13. $12 + 12 + 12 = 36"$
14. 3 dimes = 30¢
    3 nickels = 15¢
    $30¢ + 15¢ = 45¢$
    $45¢ - 5¢ = 40¢$ or $.40
15. $9 - 3 = 6$ pies

Unit Test II

1. $6 = 6$
2. $5 < 7$
3. $117 < 171$
4. 50
5. 90
6. 70
7. 100
8. 400
9. 700
10. 2, 4, 6, 8, 10, 12, 14, 16, 18, 20
11. 5, 10, 15, 20, 25, 30, 35, 40, 45, 50

12. 10, 20, 30, 40, 50, 60, 70, 80, 90, 100

13.
$$\begin{array}{r} 13 \\ +45 \\ \hline 58 \end{array}$$

14.
$$\begin{array}{r} 1 \\ 64 \\ +28 \\ \hline 92 \end{array}$$

15.
$$\begin{array}{r} 1 \\ 76 \\ +19 \\ \hline 95 \end{array}$$

16.
$$\begin{array}{r} 1\ 1 \\ \$3.6\ 1 \\ +1.79 \\ \hline \$5.40 \end{array}$$

17.
$$\begin{array}{r} 1 \\ 452 \\ +256 \\ \hline 708 \end{array}$$

18.
$$\begin{array}{r} 11 \\ 968 \\ +\ \ 75 \\ \hline 1,043 \end{array}$$

19. $7+3+2=12$

20. $5+4+5+1=15$

21.
$$\begin{array}{r} 1 \\ 65 \\ 12 \\ 18 \\ +\ \ 3 \\ \hline 98 \end{array}$$

22. rectangle
$19+21+19+21=80"$

23. $14-5=9$

24. $18-9=9$

25. $10-4=6$

26. $17-8=9$

27. $7-5=2$

28. $11-6=5$

29. $12+12+12+12=48"$

30. 3 dimes = 30¢
6 nickels = 30¢
$30¢+30¢+7¢=67¢$ or $.67

Test 16

1. 8,672
"eight thousand,
six hundred seventy-two"

2. 93,145
"ninety-three thousand,
one hundred forty-five"

3. 236,179

4. 11,416

5.
$$\begin{array}{r} 1 \\ 763 \\ +5\ 18 \\ \hline 1,28\ 1 \end{array}$$

6.
$$\begin{array}{r} 11 \\ 354 \\ +956 \\ \hline 1,3\ 10 \end{array}$$

7.
$$\begin{array}{r} 1\ \ 1 \\ \$3.56 \\ +2.49 \\ \hline \$6.0\ 5 \end{array}$$

8.
$$\begin{array}{r} 1 \\ 56 \\ 44 \\ +98 \\ \hline 198 \end{array}$$

9. $14-9=5$

10. $13-5=8$

11. $16-7=9$

12. $11-4=7$

13. $10-6=4$

14. $12-8=4$

15. 4 nickels = 20¢
3 dimes = 30¢
30¢ > 20¢

16. $4+7=11$
$11-8=3$

Test 17

1. 2,000
2. 10,000
3.
```
  1  1
  6,309
+ 1,712
  8,021
```
4.
```
      1
  8,416
+ 3,554
 11,970
```
5.
```
       1
  $ 8.92
+   4.25
  $13.17
```
6. 786,410;
   seven hundred eighty-six thousand,
   four hundred ten
7. $12+16+20 = 48'$
8. $10-3 = 7$
9. $9-5 = 4$
10. $15-6 = 9$
11. $12-4 = 8$
12. $14-7 = 7$
13. $7-4 = 3$
14. $14-5 = 9$ birds
15. $2,176+3,402 = 5,578$ ants
16. $\$29+\$21+\$9 = \$59$

Test 18

1.
```
   12
  123
  678
  207
  133
+ 312
1,453
```
2.
```
   11
  662
  108
  500
  543
+ 161
1,974
```

3.
```
   22
  782
  153
  269
  341
+ 807
2,352
```
4. 871,465
   "eight hundred seventy-one thousand,
   four hundred sixty-five"
5. 56,217
6. $15-9 = 6$
7. $10-8 = 2$
8. $13-9 = 4$
9. $5-3 = 2$
10. $17-8 = 9$
11. $18-9 = 9$
12. $7-6 = 1$
13. $11-8 = 3$
14. 2, 4, 6, 8, 10, 12, 14, 16, 18, 20
15. $5+7+4+10+13 = 39$ miles
16. $15-7 = 8$ people

Test 19

1.
```
   211
  2,834
  1,548
+ 3,672
  8,054
```
2.
```
   1 11
  4,506
  3,294
+ 2,753
 10,553
```
3.
```
   1 11
  6,729
  5,326
+ 8,361
 20,416
```
4. $10-6 = 4$
5. $16-9 = 7$
6. $6-3 = 3$
7. $11-4 = 7$

8. $9 - 5 = 4$
9. $14 - 7 = 7$
10. $11 - 6 = 5$
11. $10 - 5 = 5$
12. $1 = 1$
13. $18 > 8$
14. $209 < 902$
15. $1,367 + 2,079 + 1,534 = 4,980$ trees
16. $10 + 10 + 10 = 30"$
17. 10, 20, 30, 40, 50, 60, <u>70</u> toes
18. $\$.46 + \$.05 = \$.51$

10. 54,971; "fifty-four thousand, nine hundred seventy-one"
11. 2, 4, 6, 8, 10, 12
    14, 16, 18, 20
12. 5, 10, 15, 20, 25
    30, 35, 40, 45, 50
13. 10, 20, 30, 40, 50
    60, 70, 80, 90, 100
14. $29 - 26 = 3$
15. $48 - 34 = 14$

Test 20

1.
$$\begin{array}{r} 50 \\ -20 \\ \hline 30 \end{array} \quad \begin{array}{r} 20 \\ +30 \\ \hline 50 \end{array}$$

2.
$$\begin{array}{r} 42 \\ -30 \\ \hline 12 \end{array} \quad \begin{array}{r} 30 \\ +12 \\ \hline 42 \end{array}$$

3.
$$\begin{array}{r} 85 \\ -23 \\ \hline 62 \end{array} \quad \begin{array}{r} 23 \\ +62 \\ \hline 85 \end{array}$$

4.
$$\begin{array}{r} 263 \\ -112 \\ \hline 151 \end{array} \quad \begin{array}{r} 112 \\ +151 \\ \hline 263 \end{array}$$

5.
$$\begin{array}{r} 888 \\ -546 \\ \hline 342 \end{array} \quad \begin{array}{r} 546 \\ +342 \\ \hline 888 \end{array}$$

6.
$$\begin{array}{r} 374 \\ -62 \\ \hline 312 \end{array} \quad \begin{array}{r} 62 \\ +312 \\ \hline 374 \end{array}$$

7.
$$\begin{array}{r} 3274 \\ 1690 \\ +7216 \\ \hline 12,180 \end{array}$$

8.
$$\begin{array}{r} 5209 \\ 6128 \\ +1342 \\ \hline 12,679 \end{array}$$

9.
$$\begin{array}{r} 123 \\ 456 \\ 789 \\ +111 \\ \hline 1,479 \end{array}$$

Test 21

1. :15
2. :47
3. :36
4. :05
5.
$$\begin{array}{r} 78 \\ -6 \\ \hline 72 \end{array}$$

6.
$$\begin{array}{r} 562 \\ -41 \\ \hline 521 \end{array}$$

7.
$$\begin{array}{r} 826 \\ -603 \\ \hline 223 \end{array}$$

8.
$$\begin{array}{r} \$7.20 \\ +4.56 \\ \hline \$11.76 \end{array}$$

9.
$$\begin{array}{r} 1\,1\,1 \\ 6,034 \\ +2,987 \\ \hline 9,021 \end{array}$$

10.
$$\begin{array}{r} 2\,1 \\ 594 \\ 346 \\ +273 \\ \hline 1,213 \end{array}$$

11. $25 - 13 = 12$ years
12. $38 + 107 + 73 = 218$ miles

Test 22

1.
```
  5 1        1
  6̷2        28
 −2 8       +34
  3 4        62
```

2.
```
  2 1        1
  3̷4        15
 −1 5       +19
  1 9        34
```

3.
```
  8 1        1
  9̷3        56
 −5 6       +37
  3 7        93
```

4.
```
  1 1        1
  2̷2        18
 −1 8       + 4
    4        22
```

5.
```
  8 1 7      2 0 5
 −2 0 5     +6 1 2
  6 1 2      8 1 7
```

6.
```
  6 2 3      5 1 2
 −5 1 2     +1 1 1
  1 1 1      6 2 3
```

7. $7.33

8. 11,385

9. 1,396

10. : 40

11. : 59

12. 35 − 9 = 26;
26 − 9 = 17
OR
9 + 9 = 18;
35 − 18 = 17

Unit Test III

1.
```
  5         1
  6̷¹8      19
 −1 9      +49
  4 9       68
```

2.
```
  3         1
  4̷¹4      36
 −3 6      + 8
    8       44
```

3.
```
  7         1
  8̷¹3      58
 −5 8      +25
  2 5       83
```

4.
```
  6         1
  7̷¹2      16
 −1 6      +56
  5 6       72
```

5.
```
  1         1
  2̷¹5       9
 −  9      +16
  1 6       25
```

6.
```
  8         1
  9̷¹6      47
 −4 7      +49
  4 9       96
```

7.
```
  8 9       2 5
 −2 5      +6 4
  6 4       8 9
```

8.
```
  9 4 6      6 3 2
 −6 3 2     +3 1 4
  3 1 4      9 4 6
```

9.
```
  7 5 1      3 2 0
 −3 2 0     +4 3 1
  4 3 1      7 5 1
```

10.
```
   1 1 1
   4,826
   1,493
  +5,066
  11,385
```

11.
```
    1   1
   3,926
   3,238
  +1,221
   8,385
```

12.
```
      2
     952
     381
     381
   +  63
   1,777
```

13. 56,142; "fifty-six thousand, one hundred forty-two"

14. 224,651

15. 3,000

16. 5,000

17. : 21

18. : 50
19. 2,452 + 1,079 + 958 = 4,489 people
20. 255 + 125 = 380 dollars
     380 − 140 = 240 dollars

Test 23

1. 4 : 55
2. 7 : 15
3. 3 : 03
4. 6 : 00
5.
```
  81      1
  90     13
- 13    +77
  77     90
```
6.
```
  31      1
  44     35
- 35    + 9
   9     44
```
7.
```
  41      1
  52     29
- 29    +23
  23     52
```
8.
```
  638     117
- 117    +521
  521     638
```
9. $3.78 + $.55 = $4.33
10. 3 dimes = 30¢
     5 nickels = 25¢
     30¢ + 25¢ + 16¢ = 71¢ or $.71

Test 24

1.
```
  1 9      11
  2 0 0     38
  -  3 8  +162
    162     200
```
2.
```
    5       1
  3 6 7    149
  -1 4 9  +218
    218     367
```

3.
```
  6 14      11
  7 5 5    286
  -2 8 6  +469
    469     755
```
4.
```
  7 17      11
  8 8 0     94
  -  9 4  +786
    786     880
```
5.
```
  48      14
 -14     +34
  34      48
```
6.
```
  53      42
 -42     +11
  11      53
```
7. 7 : 09
8. 3 : 47
9.
```
    1
  650
 +163
  813
```
10.
```
    1
  274
 +895
 1,169
```
11.
```
    1
  3,039
 +5,211
  8,250
```
12. 313 − 194 = 119 pages
13. 10 + 12 + 15 = 37'
     37 − 19 = 18'
14. 3,100 + 4,916 = 8,016 mi

Test 25

1. February
2. fourth
3. twelfth
4. first
5. second
6. third
7. fourth
8. fifth
9. sixth
10. seventh

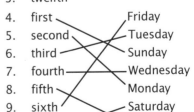

Friday
Tuesday
Sunday
Wednesday
Monday
Saturday
Thursday

11. 30
12. 31
13. 15
14. 12
15. ||||\ ||||\ ||||\ |||
16. ||||\ ||||\ ||||\ ||||\ ||||\

17.
$$\begin{array}{r} 2\,^{1}3 \\ \cancel{3}\,\cancel{4}\,^{1}1 \\ -6\,9 \\ \hline 2\,7\,2 \end{array}$$

18.
$$\begin{array}{r} 3 \\ 7\,\cancel{4}\,^{1}3 \\ -2\,0\,5 \\ \hline 5\,3\,8 \end{array}$$

19.
$$\begin{array}{r} 5\,9 \\ \cancel{6}\,\cancel{0}\,^{1}0 \\ -3\,5 \\ \hline 5\,6\,5 \end{array}$$

20.
$$\begin{array}{r} 1 \\ 5{,}1\,5\,8 \\ +1{,}2\,2\,2 \\ \hline 6{,}3\,8\,0 \end{array}$$

4.
$$\begin{array}{r} 8 \\ 4\,\cancel{9}\,^{1}3 \\ -5\,8 \\ \hline 4\,3\,5 \end{array} \qquad \begin{array}{r} 1 \\ 5\,8 \\ +4\,3\,5 \\ \hline 4\,9\,3 \end{array}$$

5.
$$\begin{array}{r} 7\,^{1}1 \\ \cancel{8}\,\cancel{2}\,^{1}4 \\ -3\,6\,7 \\ \hline 4\,5\,7 \end{array} \qquad \begin{array}{r} 1\,1 \\ 3\,6\,7 \\ +4\,5\,7 \\ \hline 8\,2\,4 \end{array}$$

6.
$$\begin{array}{r} 3 \\ \cancel{4}\,^{1}1 \\ -3\,2 \\ \hline 9 \end{array} \qquad \begin{array}{r} 1 \\ 3\,2 \\ +9 \\ \hline 4\,1 \end{array}$$

7. ||||\ ||||\
8. ||||\ ||||\ ||||\ ||||\ ||||\ ||||\ ||
9. ||||\ ||||\ ||||\ ||||\ ||||\ |||
10. 8 : 40
11. 3 : 20
12. Friday
13. second
14. $43 + 43 + 43 + 43 = 172"$
15. $5 + 8 + 13 = 26$ picked
    $26 - 18 = 8$ left

Test 26

1.
$$\begin{array}{r} 7\,^{1}0 \\ 7{,}\cancel{8}\,\cancel{1}\,^{1}0 \\ -6\,8\,1 \\ \hline 7{,}1\,2\,9 \end{array} \qquad \begin{array}{r} 1\,1 \\ 6\,8\,1 \\ +7{,}1\,2\,9 \\ \hline 7{,}8\,1\,0 \end{array}$$

2.
$$\begin{array}{r} 4\,^{1}0 \\ 5{,}\cancel{1}\,^{1}2\,3 \\ -4{,}8\,8\,2 \\ \hline 2\,4\,1 \end{array} \qquad \begin{array}{r} 1\,1 \\ 4{,}8\,8\,2 \\ +2\,4\,1 \\ \hline 5{,}1\,2\,3 \end{array}$$

3.
$$\begin{array}{r} 1 \\ \cancel{2}{,}\,^{1}1\,7\,5 \\ -1{,}5\,0\,4 \\ \hline 6\,7\,1 \end{array} \qquad \begin{array}{r} 1 \\ 1{,}5\,0\,4 \\ +6\,7\,1 \\ \hline 2{,}1\,7\,5 \end{array}$$

Test 27

1.
$$\begin{array}{r} 5 \\ \$5.\cancel{6}\,^{1}7 \\ -.2\,8 \\ \hline \$5.\,3\,9 \end{array}$$

2.
$$\begin{array}{r} 2 \\ \$\cancel{3}.\,^{1}1\,9 \\ -1.4\,2 \\ \hline \$1.\,7\,7 \end{array}$$

3.
$$\begin{array}{r} 5\,3 \\ \cancel{6}{,}\,^{1}3\,\cancel{4}\,^{1}0 \\ -3{,}5\,0\,6 \\ \hline 2{,}8\,3\,4 \end{array}$$

4.
$$\begin{array}{r} 1\,1 \\ 1\,7\,5 \\ +4\,8 \\ \hline 2\,2\,3 \end{array}$$

5.
```
    1 1
  $ 4.29
  + 6.88
  $11.17
```

6.
```
    1 1 1
    7,381
    2,479
   +5,630
   15,490
```

7. first ——————— January
8. second ╲ ╱ April
9. third ╲╳╱ May
10. fourth ╱╲ February
11. fifth ╱ ╲ March
12. sixth ——————— June
13. 2, 4, 6, 8, 10, 12, 14, 16, 18, 20
14. 𝍷𝍷𝍷𝍷 𝍷𝍷
15. $3.50 – $1.75 = $1.75

9.
```
     2 2
     124
     356
     279
    +451
   1,210
```

10.
```
     1 1
     729
     512
     283
    + 45
   1,569
```

11.
```
     1 1
     109
     460
     999
    + 31
   1,599
```

12. Monday
13. second
14. $54.98 – $21.15 = $33.83
15. 19,345 – 18,400 = 945 frogs

Test 28

1.
```
      3 ¹4 ¹2          1 1
   6 3,4 5 2        4 1,3 6 9
   – 4 1,3 6 9      + 2 2,0 8 3
     2 2,0 8 3        6 3,4 5 2
```

2.
```
      7 ¹2            1 1
   7 5,8 3 ¹0       2 2,5 3 6
   – 2 2,5 3 6      + 5 3,2 9 4
     5 3,2 9 4        7 5,8 3 0
```

3.
```
   $ 1 4.9 8
   –  6.1 2
   $  8.8 6
```

4.
```
         1
   $5 9.2 ¹0
   – 3 0.1 5
   $ 2 9.0 5
```

5.
```
     8   3
   $9 ¹1.4 ¹2
   – 1 5.1 7
   $ 7 6.2 5
```

6. 26 < 62
7. 2 < 4
8. 9 = 9

Test 29

1. 15
2. 100°

3.

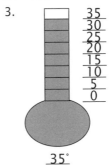

35°

4. 80
5. 30

6.
```
          1
   5 4,0 2 ¹1
   – 1 4,0 1 5
     4 0,0 0 6
```

7. $6\,{}^{1}1\,{}^{1}3$
    $\cancel{7},\cancel{2}\,\cancel{4}\,{}^{1}3$
    $-\ 3,5\ 6\ 7$
    $\overline{\ \ 3,6\ 7\ 6\ }$

8. $1\ 1$
    $9,4\ 6\ 1$
    $+2,7\ 4\ 3$
    $\overline{1\ 2,2\ 0\ 4}$

9. 1 : 25

10. 6 : 03

11. 1799 – 1732 = 67 years

12. |||| |||| |||| ||||
    |||| |||| |||| |||| |||| ||

Test 30

1. February
2. March
3. 25°
4. 40°
5. 5
6. 300°
7.

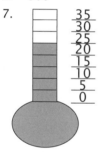

25°

8. 60
9. 10

10. $4\,{}^{1}5$
     $14,\cancel{5}\,\cancel{6}\,{}^{1}7$
     $-12,2\ 7\ 8$
     $\overline{\ \ 2,2\ 8\ 9}$

11. $3\,{}^{1}4\ 9$
     $4,\cancel{5}\,\cancel{0}\,{}^{1}0$
     $-2,6\ 9\ 9$
     $\overline{\ \ 1,8\ 0\ 1}$

12. $1\ 1\ 1$
     $4,8\ 3\ 0$
     $9,4\ 7\ 1$
     $+3,0\ 2\ 9$
     $\overline{1\ 7,3\ 3\ 0}$

Unit Test IV

1. ${}^{5}\cancel{6}\ {}^{12}\cancel{3}\ {}^{1}2$
    $-\quad 4\quad 4\ 9$
    $\overline{\quad\ \ 1\quad 8\ 3}$

2. $7\ {}^{4}\cancel{5}\ {}^{1}0$
    $-\ 5\quad 3\ 6$
    $\overline{\quad\ 2\quad 1\ 4}$

3. $7\ {}^{7}\cancel{8}\ {}^{1}9$
    $-\ \ 2\ 6\ 3$
    $\overline{\quad 5\ 5\ 6}$

4. ${}^{4}\cancel{5},{}^{11}\cancel{2}\ {}^{9}\cancel{0}\ {}^{1}8$
    $-\quad 1,\ 6\ 1\ 9$
    $\overline{\quad\ \ 3,\ 5\ 8\ 9}$

5. ${}^{8}\cancel{9}\ {}^{17}\cancel{8},{}^{12}\cancel{3}\ {}^{13}\cancel{4}\ {}^{1}2$
    $-\quad 6\ \ 8,\ 4\ \ 5\ 3$
    $\overline{\quad\ \ 2\ \ 9,\ 8\ \ 8\ 9}$

6. $\$\ {}^{6}\cancel{7}\ {}^{11}\cancel{\$}\ {}^{1}.{}^{1}4$
    $-\quad 3\ \ 4.2\ 1$
    $\overline{\quad\ \ 3\ \ 7.9\ 3}$

7. 12 : 30
8. 2 : 23
9. 25
10. 9
11. |||| |||| |||| ||||
12. |||| |||| |||| ||||
    |||| |||| |
13. Thursday
14. Tuesday
15. fourth
16. twelfth
17. Wednesday
18. 15
19. Monday and Tuesday
20. 20
21. 400°

22.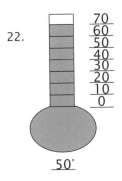

$\underline{50°}$

23. 100
24. 70

Final Test

1. 14 > 7
2. 105 < 125
3. 40
4. 70
5. 200
6. 600
7. 2,000
8. 4,000
9. 2, 4, 6, 8, 10, 12, 14, 16, 18, 20
10. 5, 10, 15, 20, 25, 30, 35, 40, 45, 50
11. 10, 20, 30, 40, 50, 60, 70, 80, 90, 100

12.
$$\begin{array}{r} {}^{1}24 \\ +46 \\ \hline 70 \end{array}$$

13.
$$\begin{array}{r} {}^{1}{}_{1}92 \\ +3\ 59 \\ \hline 5\ 51 \end{array}$$

14.
$$\begin{array}{r} 9\ {}^{1}07 \\ +1\ 68 \\ \hline 1,0\ 75 \end{array}$$

15.
$$\begin{array}{r} \$\ {}^{1}8.9 \\ +\ 2.49 \\ \hline \$11.4\ 1 \end{array}$$

16.
$$\begin{array}{r} {}^{1}6,{}^{1}474 \\ 7,6\ 10 \\ +\ 3,\ 685 \\ \hline 1\ 7,\ 769 \end{array}$$

17.
$$\begin{array}{r} {}^{1}9\,{}^{2}68 \\ 1\ 45 \\ 2\ 03 \\ +\ \ 75 \\ \hline 1,\ 3\ 9\ 1 \end{array}$$

18.
$$\begin{array}{r} {}^{1}\cancel{2}\,{}^{1}3 \\ -\ 1\ 7 \\ \hline 6 \end{array}$$

19.
$$\begin{array}{r} {}^{10}\cancel{1}\,{}^{1}5 \\ -\ \ 9\ 8 \\ \hline 1\ 7 \end{array}$$

20.
$$\begin{array}{r} {}^{3}\cancel{4}\,{}^{9}\cancel{0}\,{}^{1}3 \\ -\ 2\ 15 \\ \hline 1\ 8\ 8 \end{array}$$

21.
$$\begin{array}{r} {}^{6}\cancel{7}\,{}^{10}\cancel{1}\,{}^{1}0 \\ -\ 3\ 4\ 6 \\ \hline 3\ 6\ 4 \end{array}$$

22.
$$\begin{array}{r} 5,{}^{7}\cancel{8}\,{}^{12}\cancel{3}\,{}^{1}4 \\ -1,\ 0\ 5\ 7 \\ \hline 4,\ 7\ 7\ 7 \end{array}$$

23.
$$\begin{array}{r} {}^{7}\cancel{8}\,{}^{1}{}_{1}\,{}^{2}\cancel{3}\,{}^{11}\cancel{2}\,{}^{1}7 \\ -\ 4\ 5,\ 1\ 8\ 9 \\ \hline 3\ 6,\ 1\ 3\ 8 \end{array}$$

24. 276,591
25. 7 : 23
26. 100°
27. 12 + 12 + 12 + 12 = 48"
28. 14 + 14 + 14 + 14 = 56"
29. 6 + 11 + 6 + 11 = 34'
30. 9 + 8 + 6 = 23"

Symbols & Tables

SYMBOLS

<	less than
>	greater than
=	equals
+	plus
−	minus
¢	cents
$	dollars
°	degrees
'	foot or feet
"	inch or inches

PLACE VALUE NOTATION

$31,452 = 30,000 + 1,000 + 400 + 50 + 2$

MONTHS OF THE YEAR

January
February
March
April
May
June
July
August
September
October
November
December

MONEY AND MEASURE

1 penny = 1 cent (1¢)
1 nickel = 5 cents (5¢)
1 dime = 10 cents (10¢)
1 dollar = 100 cents (100¢ or $1.00)
12 inches (12") = 1 foot (1')
5,280 feet (ft) = 1 mile (m)
1 week = 7 days

DAYS OF THE WEEK

Sunday
Monday
Tuesday
Wednesday
Thursday
Friday
Saturday

LABELS FOR PARTS OF PROBLEMS

Addition

25	addend
+16	addend
41	sum

Subtraction

4 5	minuend
− 2 2	subtrahend
2 3	difference

Glossary

A–D

Addend - one of the numbers being added in an addition problem

Base 10 - another name for our decimal number system, which is based on 10

Borrowing - See regrouping

Cardinal Numbers - used for counting

Carrying - see Regrouping

Commutative Property - states that when adding, the order of the addends may be changed without changing the sum

Decimal Point - in this book, the dot used between dollars and cents

Decimal System - our system of numbers, which is based on 10

E–I

Equation - a number sentence where one side is equal to the other side

Estimation - used to get an approximate value of an answer

Factors - the two sides of a rectangle, or the numbers multiplied in a multiplication problem

Hundreds - the third place value in the decimal system, starting from the right

Inequality - a number sentence where one side is greater than the other side

M–P

Minuend - the first number in a subtraction problem

Minus - take away, subtract

Nineovers - a method used to check addition problems

Ordinal numbers - indicate position as first, second, etc.

Pentagon - a shape with five sides

Perimeter - the distance around a figure like a rectangle or triangle

Place value - the position of a number which tells what value it is assigned

R–S

Rectangle - a shape with four "square corners", or right angles

Regrouping - moving numbers from one place value to another in order to solve a problem. Also called "carrying" in addition, and "borrowing" in subtraction

Rounding - writing a number as its closest ten, hundred, etc., in order to estimate

Square - a rectangle with all four sides the same length. In this book, we treat squares and rectangles as two different shapes.

Subtrahend - the bottom number in a subtraction problem

T–Z

Ten Thousands - the fifth place value in the decimal system, starting from the right

Tens - the second place value in the decimal system, starting from the right

Thousands - the fourth place value in the decimal system, starting from the right. Also, the three numbers to the left of the first comma from the right in a large number

Triangle - a shape with three sides

Units - the first place value in the decimal system; also the first three numbers from the right in a large number. The word can also name measurements.

Master Index for General Math

This index lists the levels at which main topics are presented in the instruction manuals for *Primer* through *Zeta*. For more detail, see the description of each level at www.mathusee.com. (Many of these topics are also reviewed in subsequent student books.)

Beta Index